Scaling in Ecology with a Model System

MONOGRAPHS IN POPULATION BIOLOGY

SIMON A. LEVIN, ROBERT PRINGLE, AND CORINA TARNITA, SERIES EDITORS

A complete series list follows the index.

Scaling in Ecology with a Model System

Aaron M. Ellison   Nicholas J. Gotelli

PRINCETON UNIVERSITY PRESS
Princeton and Oxford

Published by Princeton University Press
41 William Street, Princeton, New Jersey 08540
6 Oxford Street, Woodstock, Oxfordshire OX20 1TR

press.princeton.edu

Library of Congress Cataloging-in-Publication Data

Names: Ellison, Aaron M., 1960– author. | Gotelli, Nicholas J., 1959– author.
Title: Scaling in ecology with a model system / Aaron M. Ellison, Nicholas J. Gotelli.
Description: Princeton : Princeton University Press, [2021] | Series: Monographs in population biology; 64 | Includes bibliographical references and index.
Identifiers: LCCN 2020049207 (print) | LCCN 2020049208 (ebook) | ISBN 9780691172705 (hardback) | ISBN 9780691222776 (paperback) | ISBN 9780691222783 (ebook)
Subjects: LCSH: Sarracenia–Research. | Biotic communities–Measurement. | Biotic communities–Statistical methods. | Botanical chemistry. | Biogeography. | Food chains (Ecology)
Classification: LCC QK495.S24 E55 2021 (print) | LCC QK495.S24 (ebook) | DDC 577.8/2–dc23
LC record available at https://lccn.loc.gov/2020049207
LC ebook record available at https://lccn.loc.gov/2020049208

British Library Cataloging-in-Publication Data is available

Editorial: Alison Kalett and Whitney Rauenhorst
Production Editorial: Kathleen Cioffi
Text and Cover Design: Carmina Alvarez
Production: Jacqueline Poirier
Publicity: Matthew Taylor and Amy Stewart
Copyeditor: Theresa Kornak

This book has been composed in Times

10 9 8 7 6 5 4 3 2 1

To our parents, Julius and Elaine Ellison and Jim and Mary Gotelli
and
To our mentors, Leo Buss, Douglass Morse, Karl Niklas, Deborah Rabinowitz,
Larry Abele, Bill Bossert, Rob Colwell, Janice Moore, Dan Simberloff,
Don Strong, Joe Travis, and Craig Young

The longer you look at one object, the more of the world you see in it.

—Flannery O'Connor (1979: 77)

Contents

Preface

Soon I'll find the right words, they'll be very simple.
—Jack Kerouac (1997: 280)

This book is a synthesis of our research on the northern (a.k.a. purple) pitcher plant *Sarracenia purpurea* viewed through the conceptual lens of "scale": scaling ecological phenomena through levels of biological organization with a model organism, across time and space, and using macroecological relationships. Before we begin, it is worth telling the story of how we started working on this most engaging plant.

In the mid-1990s, we were both newly tenured faculty members—Aaron at Mount Holyoke College in western Massachusetts, Nick at the University of Vermont after recently moving from the University of Oklahoma—with ongoing but unrelated research programs in population and community ecology, statistics, and modeling. Aaron was working on mangroves and their associated marine invertebrates in Belize (Ellison and Farnsworth 2001) and introducing ecologists to Bayesian statistics (Ellison 1996). At the same time, Nick was studying ant-lion ecology in Oklahoma (Gotelli 1997) and developing new methods of null-model analysis (Gotelli and Graves 1996). Our field work was taking each of us far from home, and we were both, independently, thinking about focusing more of our research efforts on ecological phenomena in New England.

In 1994, Aaron hired several Mount Holyoke undergraduates to help him find a new system in New England with which to study the structure and dynamic assembly of communities and food webs. Of their research, Sybil Gotch's work with the pitcher-plant system at Mount Holyoke's Hawley Bog seemed especially promising. Within a single field season, several new pitcher-plant leaves would emerge and mature, and complete food webs would begin assembling in each vase-shaped leaf as soon as it started to collect rain water. Both the plant and its food web could be monitored regularly and nondestructively and manipulated in replicated lab and field experiments. Sybil completed her senior thesis on the germination ecology of *Sarracenia purpurea* (Gotsch and Ellison 1998). The data she and the rest of the field crew collected that summer convinced Aaron that long-term experimental studies aimed at understanding community and food-web assembly in the temporally dynamic *Sarracenia* microecosystem would be feasible.

In 1995, Aaron and Nick corresponded briefly about the first (1995) edition of Nick's new textbook, *A Primer of Ecology*. We met at a poster session at the 1996 meeting of the Ecological Society of America in Providence, Rhode Island, and agreed to set up reciprocal seminar visits. In the spring of 1997, Nick visited Mount Holyoke, and Aaron pitched the idea of studying community assembly in pitcher plants. In the fall of 1997, Aaron visited the University of Vermont, and we spent a memorable hour sketching out a series of models, hypotheses, and experiments for the *Sarracenia* microecosystem. That conversation formed the road map for our first joint proposal to the US National Science Foundation (NSF). We have been collaborating ever since.

Since 1998, we have explored many facets of the genus *Sarracenia*, the *S. purpurea* microecosystem, and the larger poor-fen and bog ecosystems in which *Sarracenia* grows. Each of these projects has been driven by a general theoretical model or central questions and hypotheses that are system independent. At the same time, we have always been inspired by the unique biology and natural history of *Sarracenia*, its inquilines, its prey, its herbivores, and its place in the bog ecosystem.

We have now spent a good part of our careers focused on a single study organism whose beauty and natural history continue to enthrall us. Perhaps un-surprisingly, we were initially focused on the basic ecology and evolution of *Sarracenia*, not on the pressing problems of human impacts, which seemed far away from the isolated—and seemingly pristine—bogs of New England. But once we dug into the biological details, we discovered important influences of nutrient enrichment, habitat loss, and climatic change all around us, and we could see important connections between the biology of *Sarracenia* and the pressing envi-ronmental problems of the twenty-first century. Looking beyond *Sarracenia*, we have always found it easy to generalize from our work to other systems. In short, we continue to be motivated to "scale" *Sarracenia* and apply its lessons to other systems.

STRUCTURE

This book begins with an introductory chapter in which we discuss the importance of scale and scaling. Because ecologists use "scale" to mean many different things, it is important to be clear about how we are using these terms. The book's four main parts each consist of three chapters. In each part, the first chapter lays out a theoretical context, a basic research question, and one or more real-world appli-cations; the second chapter presents relevant observational and experimental data from (mostly) our own research into the question; and the third chapter "scales it up"—or occasionally down—to different spatial and temporal grains and different levels of biological organization. The broad themes of the four parts are (1) nutrient dynamics and stoichiometry (chapters 2–4); (2) demography and extinction (chapters 5–7); (3) community assembly and food-web dynamics (chapters 8–10); and (4) tipping points and alternative states in ecological systems (chapters 11–13). The final chapter (14) suggests a research agenda to guide future syntheses and extensions of population and community ecology. Appendices include a detailed exposition of the the the natural history of *S. purpurea* and its microecosystem; a discussion of the importance of nutrient limitation and nutrient stoichiometry; technical details on computations used in demographic, environmental niche, and tipping point models; history and methods of null models used to infer community assembly rules from co-occurrence patterns; and the potential for *omics to yield new insights into the relationship between biodiversity and ecosystem function.

AUDIENCE

We wrote this book for three kinds of readers. The first are new graduate students who are identifying study systems. We hope to convince them that it is possible to address central problems in ecology with a single organism or study system. For students, the "context" chapters that start each part may be especially valuable. In them, we present our rationales and hypotheses for studying resource acquisition and use (chapter 2), demography and population growth (chapter 5), assembly of communities and food webs (chapter 8), and tipping points (chapter 11). Detailed results and syntheses follow in subsequent chapters in each part. To enhance readability and maintain conceptual flow, more in-depth literature reviews and details of models are placed in the appendices.

The second set of readers are more established researchers who are working with different organisms and systems. We hope this group will be motivated to ask two questions. First, "how do the processes of resource acquisition, population growth, community assembly, and regime change play out in my system?" We have become convinced that the large geographical range of *S. purpurea*, its sensitivity to nutrient deposition, and its associated microecosystem and food web are ideal for studying these processes. However, it is instructive to document differences and similarities in these processes in other systems. Conversely, the second question is, "what kinds of processes are *not* easy to study with pitcher plants, but that I can address more profitably in my own research?" We love *S. purpurea*, with its unique biological and ecological quirks, but there are some disadvantages to working with this system. For example, we have benefited from the short (within-season) timescale at which we can study its food web, but *S. purpurea* itself is extremely long lived; our entire research careers have encompassed less than a single generation of this species. That makes demographic projections (Part III) very uncertain, and studying natural selection and local adaptation of this plant to rapidly changing environments is nearly impossible. All species in the American pitcher-plant family Sarraceniaceae also live in remote, fragile habitats, making for challenging field work and leading us to limit field experiments or abandon some sites to minimize our own impacts.

The third group we hope to reach includes conservation biologists, policy makers, and land managers. This might seem like a bit of a stretch coming from a pair of ivory-tower academics studying peculiar carnivorous plants growing in isolated New England bogs. Yet the problems of climatic change, nutrient enrichment, extinction risk, and regime change face all ecological systems. We hope that insights we have gained from working with *S. purpurea* can be generalized to a range of applied problems (see especially chapters 7, 10, and 13).

PERMISSIONS, DOCUMENTATION, AND REPRODUCIBILITY

The US Fish and Wildlife Service, the USDA Forest Service, the Nature Conservancy chapters in all six New England states, the Connecticut Department of Environmental Conservation, the Maine Department of Inland Fisheries and Wildlife, Katahdin Forest Management LLC, the Maine Coastal Heritage Trust, the Massachusetts Audubon Society, the Massachusetts Department of Environmental Protection, the Massachusetts Department of Fish Hatcheries, the Massachusetts Division of Conservation and Recreation, the Massachusetts Natural Heritage and Endangered Species Program, the New Hampshire Division of Resource and Economic Development, the Rhode Island Department of Environmental Protection, the Vermont Department of Fish and Wildlife, Five Colleges, Inc., Harvard Forest, the Trustees of Reservations, and the University of Vermont allowed us access to bogs and provided necessary collecting permits. Copies of all collecting permits, along with original data sheets and data notebooks, maps, and other written or printed documents are stored in the Document Archives of the Harvard Forest, located in Petersham, Massachusetts. Voucher specimens of associated fauna are accessed in the Museum of Comparative Zoology, Harvard University.

The Harvard University Botany Libraries and the Harvard University Herbaria provided scans of, and permission to reproduce, figures A.2–A.4. Robert R. F. C. Naczi and Oxford University Press permitted reproduction of figure A.1. Hannah L. Buckley and John Wiley & Sons provided permission to reproduce figure A.9. License to reproduce figure 8.1 was obtained from the Tate Museum, London. All other figures are original to this book and are copyright of the authors and

Princeton University Press, from whom permission should be obtained for future reproduction.

Permissions to reprint epigraphs found throughout the book were obtained as required; licenses to reprint copyrighted material are on file in the Harvard Forest Document Archives. Excerpt from "The Nature and Aim of Fiction" (epigraph to the book) is from MYSTERY AND MANNERS by Flannery O'Connor, edited by Sally and Robert Fitzgerald; copyright © 1962 by Flannery O'Connor, copyright renewed 1962 by Regina Cline O'Connor, and copyright © 1969 by the Estate of Mary Flannery O'Connor; reprinted by permission of Farrar, Straus and Giroux and by permission of the Mary Flannery O'Connor Charitable Trust via Harold Matson-Ben Camardi, Inc. Quotation from SOME OF THE DHARMA by Jack Kerouac (epigraph to the Preface) is copyright © 1997 by Jack Kerouac, used by permission of The Wylie Agency LLC. Excerpt from THE TENT by Margaret Atwood (epigraph to chapter 4) is copyright © 2006 O.W. Toad Ltd and reprinted by permission of McClelland & Steward, a division of Penguin Random House Canada Limited; all rights reserved; any third party use of this material, outside of this publication is prohibited; interested parties must apply directly to Penguin Random House Canada Limited for permission. Excerpt from "Thinking Against Oneself': Reflections on Cioran" (epigraph to chapter 13) is from STYLES OF RADICAL WILL by Susan Sontag. Copyright © 1967, 1969 by Susan Sontag and reprinted by permission of Farrar, Straus and Giroux.

All data presented in this book, together with the associated R code used for analysis and modeling are publicly available in digital, machine-readable form from the Harvard Forest Data Archive and the Environmental Data Initiative (see table on p. xix). The URLs and DOIs were active and accurate as of the date of publication.

ACKNOWLEDGMENTS

Although Aaron and Nick are the only named coauthors of this book, the research we discuss and synthesize is the product of field studies, laboratory projects, and intellectual collaborations done with our undergraduates, graduate students, post-docs, and many senior collaborators: Lubomír Adamec, Heidi Albright, Roxanne Ardeshiri, Lauren Ash, Dan Atwater, Vanessa Avalone, Ben Baiser, Bryan Ballif, Alana Belcon, Charles Bell, Katie Bennett, Leonora Bittleston, Leszek Błędzki, Primrose Boynton, Steve Brewer, Rachel Brooks, Hannah Buckley, Elena Butler, Jess Butler, Jonah Butler, Pat Calie, Liane Cochran-Stafira, Stefan Cover, Chuck Davis, Stephanie Day, Israel Del Toro, Philip Dixon, Bob Dorazio, Elaine Doughty, Rebecca Emerson, Elizabeth Farnsworth, Angélica González, Sybil Gotsch, Gary

Grossman, Clarisse Hart, Emily Jean Hicks, Samantha Hilerio, Abigail Hood, Natalie Hsiang, Steve Hudman, Eric Jules, Jim Karagatzides, Jamie Kneitel, Roberto Kolter, Matt Lau, Michael Lavine, Adam Maidman, Jonathan Mejia, Tom Miller, Jon Millett, Sergio Morales, Laurel Moulton, Paula Mouser, Tuyeni Mwampamba, Rob Naczi, Gidi Ne'eman, Rina Ne'eman, Amanda Northrop, Calley Ordoyne, Wyatt Oswald, Jerelyn Parker, Manisha Patel, Celeste Peterson, Lindsey Pett, Naomi Pierce, Uma Pinninti, Ann Pringle, Jamie Ratchford, Sydne Record, Relena Ribbons, Don Ross, Nate Sanders, Justine Sears, Jennie Sirota, Aidan Smith, Hedda Steinhoff, Matthew Toomey, Amy Wakefield, Sarah Wittman, Joseph Wonsil, Ben Wolfe, Anne Worley, and Regino Zamora. To all we are most grateful.

We received generous help from Judy Warnement, librarian of the Harvard University Botany Libraries, for assistance in tracking down the earliest references to, and images of, *Sarracenia purpurea*. Michelle Nijhuis's essay "On the trail of climate" published in the *New York Review of Books* suggested the epigraph to chapter 1 and Michelle herself pointed us toward the original source.

We also thank Simon Levin, Henry Horn, and Alison Kalett for encouraging us to write this book and taking it on in the Monographs in Population Biology series; David Foster, Simon Levin, Rob Pringle, Sydne Record, Corina Tarnita, and two anonymous reviewers who provided excellent feedback on the entire book; and Alison Kalett and the production team at Princeton University Press for seeing it through to publication.

We are most grateful to our home institutions—Mount Holyoke College, Harvard University, and the University of Vermont—for gainful employment, state-of-the-art facilities, stimulating intellectual environments, and a continual flow of outstanding students. This work would not have been possible without the support we have had from the NSF since 1998 from the US National Science Foundation's Division of Environmental Biology (core programs) and the Division of Biological Infrastructure's Research Experiences for Undergraduates and Field Stations & Marine Laboratories programs; the Howard Hughes Medical Institute; the Ellen P. Reese Fund at Mount Holyoke College; the Massachusetts Natural Heritage and Endangered Species Program; the Massachusetts Audubon Society; the Trustees of Reservations; the Vermont Genetics Network; and the Nantucket Biodiversity Initiative.

Finally, we would be remiss if we did not acknowledge how much easier it is to communicate our love for science to a broad audience because we work on charismatic carnivorous plants. Although *Sarracenia* is far removed phylogenetically from Audrey Jr.—Griffith's (1960) fictitious cross between a Venus' fly-trap and a bladderwort—the ability of carnivorous plants to enchant all children and children-at-heart transcends their taxonomy. Everyone can enjoy a good story; we hope you enjoy ours.

TABLE: Datasets and code for all analyses presented in the book.

Description	Chapters	Archive number[a]	DOI[b]
Prey spectra of carnivorous plants	A.1	HF111	doi: 10.6073/pasta/cb95637eda0f96c3fdbd1a97e632c7b7
S. purpurea morphology and associated ants	A.1, 3	HF159	doi: 10.6073/pasta/37ee70bbf38b3f95bc955dc555c63188
Prey additions and Sarracenia physiology	3, 4	HF109	doi: 10.6073/pasta/f21842bb8875c40da89dc8d0140ac550
S. purpurea seed dispersal and seedling establishment	A.1, 6	HF331	doi: 10.6073/pasta/abc185ad45fb56a25db1d56020d4dd25
S. purpurea morphology and cohort demographic data	A.1, 3, 6	HF202	doi: 10.6073/pasta/6486af2f63b74013342f62c42ac90d9ea
Prey capture by S. purpurea	A.1, 9, 10	HF114	doi: 10.6073/pasta/50bb20fcbc08c97dcc055d6e26ac812e
Ecophysiology of carnivorous plants	3, 4	HF168	doi: 10.6073/pasta/a0464087892ef0e7c18e6479b78d8264
Carnivorous Plant Trait Database	3, 4	TRY72	url: https://www.try-db.org/TryWeb/Data.php#72
Nitrogen deposition and S. purpurea pitcher morpology	3, 4	HF330	doi: 10.6073/pasta/c7cc14cc6a12d230390d0d2637748 6ec
Nitrogen cycling in S. purpurea	3, 4, 9	HF096	doi: 10.6073/pasta/4bdb96e0a36a0b2ead0fde855e9bce25
Allochthonous nutrients in the S. purpurea microecosystem	3, 9, 12	HF098	doi: 10.6073/pasta/003cb42be865a7302904dc4a4b6a46c1
Bog plant and pore-water stoichiometry	3, 4	HF329	doi: 10.6073/pasta/3f645303464 52fac501a17c26883e964
Construction costs of carnivorous plants	4	HF112	doi: 10.6073/pasta/d035c83c2e8d4e50b7a1e856b3f79cd1
Nitrogen uptake by S. purpurea	4	HF146	doi: 10.6073/pasta/1b260e5f47725c2e513b99f2e2d647407
Effects of prey addition on S. purpurea stoichiometry	4	HF328	doi: 10.6073/pasta/6591267269a71eadc859e2497e6643cf
Leaf traits of Darlingtonia californica	4	HF327	doi: 10.6073/pasta/5b90c085d259af1eed679a1c1bfa4a99
Occurrence records of S. purpurea in North America	7	HF332	doi: 10.6073/pasta/8b0b710ea267f23a22738b9f5e3fec6f
Macrobial food web of S. purpurea	9, 10	HF193	doi: 10.6073/pasta/3bd2c920cf431442d34227f3f5fc7c2a
Decomposition dynamics by the S. purpurea food web	9, 10	HF169	doi: 10.6073/pasta/a0464087892ef0e7c18e6479b78d8264
Metaproteome of S. purpurea	12, 13	HF295	doi: 10.6073/pasta/bb2a896d2c7e3e6d6f3f2129a6afc4ac
Tipping points in the S. purpurea microecosystem	12, 13	HF205	doi: 10.6073/pasta/ed0f0a74556bf919a80eca6e439bdc78
Hysteresis in the S. purpurea microecosystem	12, 13	HF334	doi: 10.6073/pasta/ebf6b175a6f6e44d3e9747c13f0d376c
R code for all analysis	3–13	HF340	doi: 10.6073/pasta/2483bcba241e33a088960292e483483e
Docker Virtual Machine to replicate all analyses	3–13	HF349	doi: 10.6073/pasta/5dc19f7d3cc4db6fb86a2984fe126f4a

[a] Refers to the archive number in the Harvard Forest Data Archive: https://harvardforest.fas.harvard.edu/ or the TRY Plant Trait Database: https://www.try-db.org/ (neither are version-controlled)

[b] Assigned and hosted by the Environmental Data Initiative https://environmentaldatainitiative.org/ (version-controlled)

Abbreviations

Abbreviation	Meaning	Units
[ ]	Concentration	g^{-1} or L^{-1}
A	Network ascendency	flux (e.g., time^{-1})
A	Demographic transition matrix	entries $a_{ij} = P(n_i \rightarrow n_j)$
A_{area}	Area-based photosynthetic rate	μmol $CO_2 \cdot m^{-2} \cdot s^{-1}$
A_{mass}	Mass-based photosynthetic rate	μmol or nmol $CO_2 \cdot g^{-1} \cdot s^{-1}$
ATP	Adenosine triphosphate	—
BSA	Bovine serum albumin	—
C	Connectance of a network	individual^{-1}
C	Carbon	—
CO_2	Carbon dioxide	—
DIN	Dissolved inorganic nitrogen	—
K	Potassium	—
K_{mass}	Mass-based potassium concentration	mg g^{-1}
KEGG	Kyoto Encyclopedia of Genes and Genomes	—
kya	Thousands of years (ago)	years
λ	Population growth rate	—
L	Number of edges or links in a network	individuals
LD	Average link density in a network	—
LD_e	Effective link density in a network	—
LMA	Leaf mass area	g m^{-2}
LTRE	Life-table response experiment	—
Mya	Millions of years (ago)	years
n	Sample size or number of individuals	individuals
n	Column vector in a demographic model	entries n_{ij}
N	Number of nodes in a network	individuals
N	Nitrogen	—
NADP	National Atmospheric Deposition Program	
N_{mass}	Mass-based nitrogen concentration	mg g^{-1}
NH_4	Ammonium	—
NO_2	Nitrite	—
NO_3	Nitrate	—
NO_x	Nitrous oxides	—
OTU	Operational taxonomic unit	—

P	Phosphorus	—
P_{mass}	Mass-based phosphorus concentration	$mg\ g^{-1}$
PAR	Photosynthetically active radiation	$\mu mol \cdot m^{-2} \cdot s^{-1}$
PNUE	Photosynthetic nutrient-use efficiency	$nmol\ CO_2\ mg\ N^{-1}\ s^{-1}$
PO_4	Phosphate	—
POM	Particulate organic matter	—
r	Per capita rate of population growth	$time^{-1}$
rRNA	Ribosomal RNA	—
S	Number of species	individuals
S	Sulfur	—
SDM	Species distribution model	—
SLA	Specific leaf area	$mm^2\ mg^{-1}$
T-RFLP	Terminal restriction fragment-length polymorphisms	—
TD	Trophic depth of a network	—
TD_E	Effective trophic depth of a network	—

Scaling in Ecology with a Model System

Frontispiece. Flowers of *S. purpurea* at Hawley Bog. Photograph
© Aaron M. Ellison and used with permission.

CHAPTER ONE

Introduction

Why Scale?

Es gibt überall einen Makroksomos und einen Mikrokosmos,
eine Welt im Grossen und eine Welt im Kleinen, diese eben
so wichtig, oft wichtiger als jene.[1]
—Carl Kreil (1865: 2–3)

Scaling is central to the human condition. There are literally dozens of definitions of scale derived from homonyms and cognates rooted in Scandinavian, Germanic, and Romantic languages (OED Online 2018b). These definitions include seven distinct roots and meanings of scale as a noun and three as a verb. All of these variants have subsidiary definitions, ranging from bowls, husks, huts, and balances, through horny coverings of insects, fishes, and plants, to ladders and the *scala naturae*.

Across science and philosophy, scaling in the sense of inductive reasoning or generalized modeling has an equally long history.[2] Levins' (1966) identification of trade-offs in models between generality, realism, and precision continues to inform and bedevil how we understand the world and forecast changes to it (Weisberg 2013).

Levin (1992: 1943) asserted that "the problem of pattern and scale is the central problem in ecology, unifying population biology and ecosystem science, and marrying basic and applied ecology." Ecological definitions of scale seem to be derived from the Latin *scāla* (a ladder or stairway), which entered English usage in the seventeenth or eighteenth century, and most ecological uses of scale reflect a "common-sense" descriptor of spatial (big or small), or temporal (long or short) dimension (Ellison 2018). Despite this common etymology, Levin (1992) argued that there is no single natural scale at which ecological systems or phenomena should be studied because different mechanisms operate at different spatial, temporal, biological, or organizational scales (figure 1.1; cf. figure 2 of Levin 1992). Levin also noted that the spatial or temporal scale of an analysis rarely is motivated by a particular mechanism but more typically reflects an investigator's perceptual bias or sampling constraints (i.e., the scale at which we choose to observe a system). Other ecologists have turned away from Levin's scale-specific view of nature, instead searching for scale-free patterns and processes that operate in (nearly)

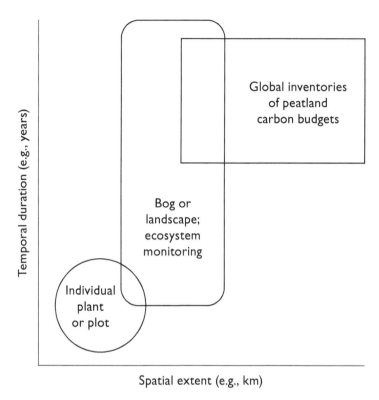

Figure 1.1. Different ecological phenomena are measured most commonly at different spatial and temporal scales. Levin's approach to scaling can be represented on a graph with two axes and a third represented by contours of studies, responses, or levels of inference (Estes et al. 2018).

identical ways at all spatial and temporal scales. Ecophysiologists, biomechanists, and macroecologists, among others, have followed the lead of physical scientists who think of scale in terms of dimensionless ratios that characterize processes or mechanisms.

1.1 TIME AND SPACE

We have adopted these and other approaches to scaling *Sarracenia*. Like most ecologists, we have implicitly adopted Levin's approach to scaling. We have studied ecophysiological processes of individual plants measured from seconds (e.g., photosynthesis: Ellison and Gotelli 2002; Ellison and Farnsworth 2005) to seasons (e.g., nutrient uptake, storage, and retranslocation: chapter 3; Ellison and Gotelli 2002; Wakefield et al. 2005; Ellison 2006; Butler and Ellison 2007; Butler et al. 2008; Karagatzides and Ellison 2009; Karagatzides et al. 2009). We have documented and modeled demographic patterns of populations from seasons to decades within and among the bogs where *Sarracenia* grows (chapter 6; Gotelli and Ellison 2002, 2006b; Ne'eman et al. 2006). We have studied the complex food webs that live within *Sarracenia* pitchers (chapter 9) at spatial scales ranging from individual pitchers (Gotelli and Ellison 2006a; Butler et al. 2008; Peterson et al. 2008; Gotelli et al. 2011) to the North American continent (Buckley et al. 2003, 2004, 2010; Baiser et al. 2011, 2013), and the temporal scales of their formation (Lau et al. 2018) (figure 1.2). When discussing the "*Sarracenia* microecosystem" in the remainder of this book, we draw on data collected from within the entire *S. purpurea* complex (including *S. rosea*; see appendix A, §A.1.3). We specify geographical, subspecific, or varietal distinctions only when these could help to explain observed patterns.

As we moved our research scale from the lower left to the upper right of figure 1.1, we also traversed the different ecological subdisciplines that typically represent these spatial and temporal scales: physiological and population ecology, community ecology, ecosystem ecology, landscape ecology, macroecology, and biogeography. Each of these subdisciplines has its own research cultures, journals, and meetings, and they interact less often than might be expected. These separate scales of inference and distinct subdisciplines reinforce Levin's argument that there is no single or natural process, equation(s), or dimensionless number(s) with which to examine ecological phenomena or at which ecological processes operate. However, there might be characteristic equations within each of these scales that could be coupled across scales (§1.3).

1.2 GENES TO ECOSYSTEMS

An alternative interpretation of scale describes the hierarchical levels of biological organization: from genes and genomes to organisms, populations, and ecosystems. This scale has both parallels and some overlap with the time-and-space scales of figure 1.1. The boundaries set by different scales of biological organization also are manifest in the structuring of academic departments and funding agencies. These structural constraints, along with differences in the technical skills, language, and conceptual frameworks (e.g., reductionism, holism, emergence), have made it challenging to explore and transcend boundaries of biological scales. Mechanistic explanations that cross biological scales often move in a linear reductionist chain (e.g., selfish genes drive individual behavior, which in turn affects population-level dynamics). However, our work has helped us to understand patterns and processes that jump scales in complex sequences. For example, nutrient inputs

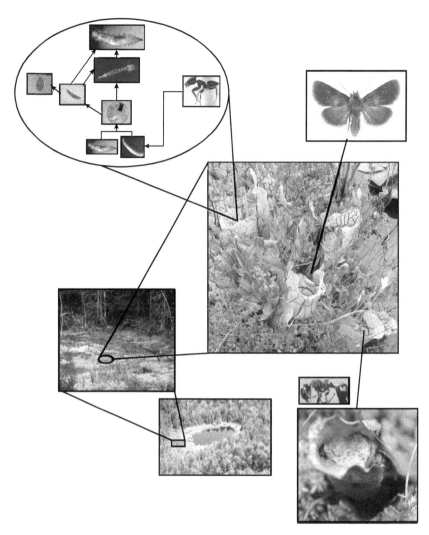

FIGURE 1.2. The *Sarracenia purpurea* microecosystem includes pitcher plants (center); an aquatic, detritus-based food web within each pitcher (top left); moth larvae that feed on rhizomes, pitchers, and seeds (top right); and ants that nest in old pitchers (bottom right) that are the primary prey of the plant (and the detrital input for the food web). Pitcher plants occur in individual bogs, which are isolated features of larger landscapes (bottom left). Original photographs and photomontage ©Aaron M. Ellison and used with permission.

affect pitcher-plant growth (chapter 3), which not only affects demography and population persistence (chapters 6 and 7) but also directly influences food-web development (chapter 9) and alternative ecosystem states (chapters 12 and 13).

1.3 MODELING: METABOLIC THEORY AND MACROECOLOGY

Ecological modeling provides other ways to cross scales (figure 1.3). Although some models capture processes and dynamics of multiple scales of space, time, or biological organization, there is as yet no unified ecological process model that encompasses all of these scales. Metabolic theory and its extensions into macroecology offer an important alternative to scale-specific frameworks and a first step toward a unified model of ecological processes (e.g., West et al. 1997; Enquist et al. 2003; Brown et al. 2004; Shade et al. 2018). The emphasis in macroecology has been on measuring one or more aspects of organismal body size and exploring the functional and mathematical relationships among them. Such dimensional analyses might yield a general process equation for multiple spatio-temporal scales (figure 1.1) or provide links to the processes operating between different scales (figure 1.3).

Although the starting point of West et al. (1997) was the assertion that biological diversity is essentially a reflection of body size (mass), they subsequently used the general allometric (i.e., power law) equation, $Y = Y_0 M^b$, to relate body mass (M) to a diversity of physiological and metabolic variables (Y), including metabolic rate, lifespan, cross-sectional area of vessels, and the like. Macroecologists have now documented many examples of allometric scaling of ecological phenomena. Of particular interest are systemic differences between sublinear scaling ($b < 1$ in the power-law equation) and superlinear scaling ($b > 1$). Phenomena that scale sublinearly show efficiencies of scale, and many biological scaling relationships are sublinear (e.g., the global spectrum of leaf traits, Wright et al. 2004, and see chapter 4). In contrast, phenomena that scale superlinearly are more productive at larger scales. Examples of superlinear scaling relationships are rare—West (2017) found them mostly in cities, companies, and other human organizations—but they may be emergent properties of particular kinds of networks, including ecological ones (Zhang et al. 2015, and see chapters 9 and 10). The "Holy Grail" of metabolic theory and macroecology is the identification of a single underlying mechanism for the wide range of ecological patterns characterized by scaling laws. Common patterns such as quarter-power scaling for physiological processes suggest common mechanisms, but Levin's observation that "correlations are no substitute for mechanistic understanding" (Levin 1992: 1960) bears repeating.

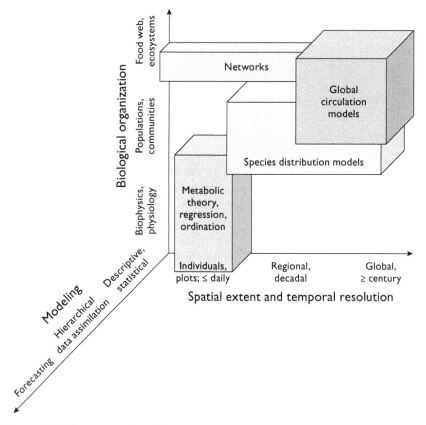

1.4 MECHANISMS AT SCALES

By defining function in terms of relationships between levels of organization, Farnsworth et al. (2017) highlighted the role of interactions—and the importance of networks of interactions—as a proximate cause (i.e., a one-level-lower mechanism) of many ecological phenomena. They suggested that experimental manipulations that distinguish effects of network structure (emergent, community-level effects) from effects of traits (organism-level effects) could provide new insights into emergent properties of ecological communities, entire ecosystems, and their relationships (Barry et al. 2019). Our experimental manipulations of the *Sarracenia* microecosystem (chapters 9 and 12) illustrate the power of this approach.

Our studies of tipping points in the *Sarracenia* microecosystem (chapters 12 and 13; Sirota et al. 2013; Northrop et al. 2017; Lau et al. 2018) explicitly take advantage of dimensionless (i.e., scale-independent) variables. We have also used comparative studies of carnivorous plants, including species in the genus *Sarracenia* (Farnsworth and Ellison 2008), to explore taxon-specific divergence from the macroecological scaling laws for leaf traits (chapter 4; West et al. 1997; Wright et al. 2004). The deviations from general scaling exponents detected in these and other counterexamples tend to disappear, or at least regress to the mean, when aggregated data are plotted on a double-log scale.

1.5 ORGANISMS AS MODEL SYSTEMS

A final, perhaps less-intuitive, approach to scaling is the use of model systems in biological research. A handful of species—the bacterium *Escherichia coli* (Migula), brewer's yeast *Saccharomyces cervisea* Meyen *ex* E. C. Hansen, the cellular slime mold *Dictyostelium discoidium* Raper, the roundworm *Caenorhabditis elegans* (Maupas), the fruit fly *Drosophila melanogaster* Meigen, the house mouse *Mus muculus* L., and the mouse-eared cress *Arabidopsis thaliana* (L.) Heynh.—have been developed as "model organisms" by geneticists and molecular, cell, and developmental biologists. These species were not chosen for particular properties. Rather, their utility for answering a broad range of questions that span levels of biological organization emerged over time for a variety of reasons: ease of breeding, rearing, and cultivation in the laboratory; short generation times; and early attention to their underlying genotypes (and later, genomes), genetic mutations, and phenotypic variants that could be maintained in stock cultures. Each of these model organisms also had a "champion," an individual scientist who, along with their students and colleagues, developed research networks based on the species and encouraged the sharing of cultures, genotypes, phenotypes, and data to address theoretical questions of general interest (Fields and Johnston 2005).

We nominate *Sarracenia purpurea* as a model ecological species (see appendix A for the fine-grained details of this remarkable plant) but recognize that we are bucking the ecological tradition of eschewing model organisms. Most ecologists are drawn to experimental or observational work in the field, but model organisms have been bred and optimized for work done in laboratories, growth chambers, or greenhouses and are difficult to find in the field (cf. Buss 1982; Ellison and Buss 1983; Ellison et al. 1994). With the exception of *Drosophila*, relatively little is known of the basic natural history of model organisms, their geographical distribution, or their interactions with other organisms (Yu et al. 2016). Ecologists also have had a social culture that, until recently, has rewarded individuals who study distinctive taxa in more-or-less pristine sites. However, that culture is changing: twenty-first-century ecology is becoming more a product of large collaborative groups than of solitary individuals working at their field sites. Many of these collaborative groups collate, synthesize, and analyze existing data, whereas existing and emergent networks of field studies and sites tend to be focused on common processes, not common taxa or model organisms.

Compared to other biological disciplines, we think that ecology has suffered for want of explicit model organisms and model systems. Nevertheless, a number of taxa and systems have been studied by evolutionary ecologists for many years, emphasizing processes and dynamics at specific scales. Textbook examples include decades of work on *Mytilus californianus* Conrad and *Pisaster ochraceus* (Brandt) on Tatoosh Island in Puget Sound that focused on trophic dynamics and the role of keystone species (e.g., Paine 2002; Wootton 2010; Logan et al. 2012); long-term studies of phenology and pollination in alpine meadows at Rocky Mountain Biological Laboratory (e.g., CarraDonna and Inouye 2015; Wright et al. 2015); detailed studies of social behavior of *Pogonomyrmex barbatus* Smith ants in Arizona (e.g., Gordon 2010); manipulations of grasses, grasshoppers, and their predators in food webs of northeastern old fields (Schmitz 1994, 2003, 2010); studies of goldenrods, gallmakers, and their natural enemies (Abrahamson and Weis 1997); and of course, food-web dynamics of the inhabitants ("inquilines") of the water-filled leaves and bracts ("phytotelmata") of *Sarracenia purpurea* L., *Nepenthes* Dumort. spp., and a variety of bromeliads (e.g., Frank and Lounibos 1983; Kitching 2000; Ellison et al. 2003; Srivastava et al. 2004). We hope that this book, with our syntheses of general theories informed by the detailed natural history knowledge of a small number of intimately interacting taxa, makes a convincing case that the *Sarracenia* microecosystem is an excellent model system for ecological research at a variety of scales, however they may be construed.

1.6 SUMMARY

"Scale" has been a defining concept of ecological research for nearly 30 years, but it is used in many different ways. Most ecologists use a common-sense definition of scale to mean the extent and grain of spatial and temporal phenomena, and aspects of size, morphology, and life history of organisms. These standard usages of scale intersect with different levels of biological organization to provide a framework through which ecological phenomena can be viewed. None of these different scales of study or observation has primacy, but different processes operating at different scales may be linked by conceptual models or process equations. An alternative scaling framework uses dimensionless ratios to characterize processes or mechanisms independent of space or time. This approach to scaling is characteristic of the metabolic theory of ecology and macroecology. We use all these ways of scaling to synthesize our work with the *Sarracenia* microecosystem and champion its use as a model system for studying pressing basic and applied ecological questions.

Part I

Ecophysiology, Nutrient Limitation, and Stoichiometry

CHAPTER TWO

Context

Nutrient Limitation, the Evolution of Botanical Carnivory, and Environmental Change

You are what you eat (plus a few ‰).
—Michael J. DeNiro and Samuel Epstein (1976: 834)

How do organisms acquire and use resources? These are fundamental ecological questions, and the answers have broad implications for any approach to scaling. First, resources are distributed unequally in time and space, and strategies for acquiring them may change with spatiotemporal scale. Second, resource use is a metabolic process that involves genetic and cellular regulatory networks and interacting organs to deliver nutrients to the organism. These processes have been scaled to entire ecosystems, where fluxes of energy and nutrients are modeled analogously to metabolic processes of single organisms (e.g., Patten and Odum 1981). In a parallel vein, the metabolic theory of ecology uses dimensional analysis of resource distribution to predict relationships between metabolic rate, body size, population size, diversity, and ecosystem function (West et al. 1997; Brown et al. 2004; West 2017; Barry et al. 2019). Our studies of nutrient acquisition, allocation, and use by *Sarracenia purpurea* illustrate these different aspects of scaling.

Plants and animals have different strategies for acquiring resources (Ellison 2020), which can be limited in an absolute sense (e.g., light, space, and water) or in a relative sense (e.g., nitrogen, phosphorus, and other macro- and micronutrients; see appendix B). Both types of resource limitation can be analyzed in terms of costs and benefits (Givnish 1986), which have been strong selective forces molding the evolution of suites of botanical traits (e.g., Givnish and Vermeij 1976; Reich 2014) and constraining scaling relationships between them (Wright et al. 2004; Díaz et al. 2016). These constraints also are central to the evolution, expression, and plasticity of botanical carnivory and the response of *S. purpurea* to changes in nutrient availability from local to continental scales (Givnish et al. 1984; Ellison and Gotelli 2002; Ellison 2006; Fleischmann et al. 2018).

Chapter 3 details these local-scale effects by contrasting responses of *S. purpurea* and other phytotelmata to manipulations of nutrient availability from natural (i.e., prey) and anthropogenic (i.e., atmospheric deposition) sources. These patterns and processes are "scaled up" in chapter 4, where we map the results of nutrient addition experiments with *S. purpurea* and other carnivorous plants onto the broad spectrum of metabolic and morphological leaf-trait space (Wright et al. 2004). We conclude that, in the absence of anthropogenic inputs, carnivorous plants generally follow the "rules" of nutrient stoichiometry derived from studies of noncarnivorous plants. However, *S. purpurea* and other carnivorous plants break these rules in anthropogenically modified environments.

2.1 BACKGROUND

2.1.1 Nutrient Acquisition, Plant Traits, and the Evolution
of Botanical Carnivory

Evolutionary solutions to the ecological problem of acquiring, using, and allocating limiting resources have resulted in only a limited palette of physiological and morphological traits of leaves and roots (figure 2.1; Wright et al. 2004). For vascular plants as a whole, much of this interspecific variation in functional traits also can be mapped onto two major axes: an axis of body and seed size (ranging from small to large) and an axis of leaf "construction costs" (ranging from cheap, fragile, highly productive, and palatable leaves to costly, robust, low-productivity, and unpalatable leaves; Reich 2014; Díaz et al. 2016).

Givnish and Vermeij (1976) and Givnish (1986) modeled the evolution of plant traits as the solution to a cost–benefit model that maximizes marginal photo-synthetic gain and minimizes the marginal energetic costs of constructing and maintaining leaves, shoots, and roots. Givnish et al. (1984, 2018) applied this general model to the evolution of carnivorous plants (figure 2.2). Because of the high costs of constructing plant traps and the constraint of living in very low N environments, the placement of carnivorous plants in the trait spectrum pro-vides instructive examples and counterexamples of the overall trends for vascular plants (chapter 4; Ellison and Farnsworth 2005; Farnsworth and Ellison 2008; Karagatzides and Ellison 2009).

2.1.2 Anthropogenic Activities Alter Resource Availability and Fluxes

Contemporary cost–benefit models make predictions about form and function of whole plants and individual plant organs—spectra of plant traits—based on a combination of light availability, nutrient and water limitation, and stoichiometry. These models assume that the environmental conditions in the habitats where

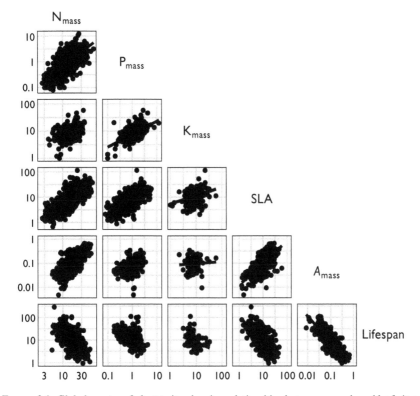

FIGURE 2.1. Global spectra of plant traits, showing relationships between mass-based leaf nitrogen (N_{mass}), phosphorus (P_{mass}), and potassium (K_{mass}) content (all in mg/g); specific leaf area (in mm^2/mg); mass-based maximum photosynthetic rate (A_{mass} in $\mu mol \cdot g^{-1} \cdot s^{-1}$); and leaf lifespan (in months). Data compiled from the TRY database (Kattge et al. 2020; details of the datasets we used are discussed below and in chapter 5).

specific plant traits evolved were roughly constant for many generations, allowing stabilizing selection to fine-tune plant morphology and physiology for a wide range of plant species. But anthropogenic forcing factors, including changes in atmospheric carbon dioxide concentrations and nutrient deposition, have changed the global environment and altered the habitats in which plants live (figure 2.3).

The US Clean Air Act (42 USC. §7401 et seq.) has, in the short term, reduced the deposition of N- (and S-) containing compounds ("acid rain") that damage plants and alter stoichiometric ratios, nutrient fluxes, and ecosystem dynamics (e.g., Warren et al. 2017; Greenbaum 2018). Mahowald et al. (2008) estimated that 5–15% of total P and PO_4-P is derived from anthropogenic sources; P from agricultural fertilizers, desert dusts, and fine soils also is a small but measurable

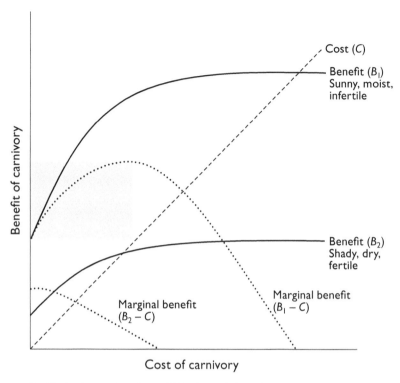

FIGURE 2.2. The basic cost–benefit model for the evolution of botanical carnivory reflects trade-offs between costs of producing traps and benefits accruing from them. Costs are assumed to increase constantly with allocation to traps, whereas benefits increase and then level off when plants reach their photosynthetic maxima. Carnivory is expected to evolve whenever the marginal benefits of additional investments in carnivorous structures exceed the marginal costs of doing so (i.e., within the gray shaded area, where the slope of the benefits–cost curve is positive). This occurs on the upward limb of the B curve, when C is small and $dB/dx > dC/dx$. These conditions are most likely in sunny, moist, and nutrient-poor habitats. Modified from Givnish et al. (1984, 2018).

component of atmospheric deposition (Allen et al. 1968; Gore 1968; Boynton et al. 1995; Ahn and James 2001; Mladenov et al. 2012; Tipping et al. 2014). Because P is rapidly taken up and reused in many ecosystems and can increase primary productivity in P-limited systems (Benitez-Nelson 2000; Paytan and McLaughlin 2007), the effects of atmospherically deposited P may be larger than previously appreciated (Mahowald et al. 2008; Tipping et al. 2014). However, the large amount of atmospheric N deposition has resulted in stoichiometric imbalance of N and P—and hence, N limitation—so P deposition may be of some importance in estimating environmentally driven N limitation. We explore the interaction

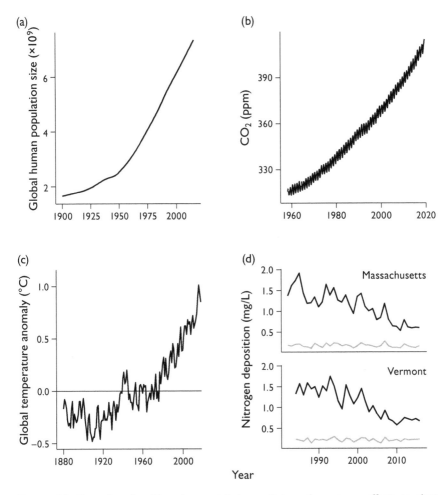

FIGURE 2.3. Examples of rapid environmental change that can have strong effects on plant ecophysiology or morphology and that we have addressed in our studies with *S. purpurea*. (**a**) Global human population growth since 1900. (**b**) Atmospheric [CO_2] measured at the Mauna Loa (Hawai'i) observatory since 1958. (**c**) Global temperature anomalies since 1880. The baseline of 0 is the average global temperature from 1951–1980. (**d**) Atmospheric deposition (mg/L) of NO_3 (black line) and NH_4 (gray line) at National Atmospheric Deposition Program (NADP) monitoring stations in Quabbin, Massachusetts (1982–2016) and Underhill, Vermont (1984–2016). *Sources:* (a) https://ourworldindata.org/world-population-growth; (b) http://scrippsco2.ucsd.edu/data/atmospheric_co2/primary_mlo_co2_record; (c) https://climate.nasa.gov/vital-signs/global-temperature/; (d) https://nadp.slh.wisc.edu/data/

between N deposition and P limitation in more detail for *S. purpurea* in chapters 3 (§3.1.2, §3.2), 4 (§4.2), and 7 (§7.2.2, 7.2.3).

All these environmental changes have happened very rapidly, over a period of only decades or a few centuries. The morphology, physiology, and phenology of many annual plants have evolved in tandem with these environmental changes (e.g., Chapin et al. 1993; Ackerly et al. 2000; Franks et al. 2007; Anderson et al. 2012a,b; Sardens and Peñuelas 2012). In contrast, many of these environmental changes have happened within only a few generations of long-lived perennial plants such as *Sarracenia*. Thus, most of the observed changes in the morphology and growth form of perennial plants in novel environments reflect phenotypic plasticity and morphological or physiological acclimation (e.g., Anderson et al. 2012a,b; Murren et al. 2015). It remains an open question how plasticity and acclimation can be mapped onto global spectra of plant traits. These spectra illustrate interspecific patterns but collapse intraspecific variation into a single, mean value. Our work with *S. purpurea* and other carnivorous plants illustrates that intraspecific spectra of leaf traits do not always match expected interspecific patterns (chapter 4; see also Zaoli et al. 2017; Fajardo and Siefert 2018; Osnas et al. 2018). This intraspecific variation in *S. purpurea* morphology also has made it a useful indicator of atmospheric N deposition (§3.1.1; Ellison and Gotelli 2002).

2.2 NEXT STEPS

Biophysical constraints on ecophysiological processes and stoichiometry interact to yield, via evolutionary dynamics, a finite set of physiological and morphological solutions to meet these needs. These solutions reflect trade-offs and constraints among competing pressures to maximize photosynthetic rates; minimize water losses from evapotranspiration and tissue losses from herbivory; and optimize nutrient acquisition, use, and allocation. Global spectra of plant traits explain much of the interspecific variation in these trade-offs and constraints. A fixed spectrum of traits implies that the environments in which plants evolved have been relatively stable across multiple generations. However, ongoing and rapid anthropogenic environmental changes—including increase in atmospheric [CO_2], appropriation by humans of large amounts of fresh water and nutrients, and alteration of water and nutrient cycles—are occurring within a few generations of long-lived perennial taxa such as *S. purpurea*. In the short term, perennial plants can acclimate to these changes, often with plastic responses in traits and shifts in the stoichiometry of plant tissues. In chapter 3, we describe experiments and observations illustrating such plasticity and acclimation by *S. purpurea* and other carnivorous plants to variation in nutrient supply. In the long term, adaptation to chronic environmental change will require sufficient genetic variation in plant traits to respond to natural selection. In chapter 4, we consider how the "fit" of *S. purpurea* and other carnivorous plants in global spectra of plant traits provides new insights into possible longer-term responses by *S. purpurea* to ongoing environmental change.

CHAPTER THREE

The Small World

Stoichiometry and Nutrient Limitation in Pitcher Plants and Other Phytotelmata

Feed me more!
—[Audrey] Junior, in Griffith (1960: 32)

Sarracenia purpurea is an excellent system for understanding the effects of absolute and relative increases in nutrients because its nutrient uptake system is very simple. The leaves form simple cups that fill with rainwater and act as traps for arthropod prey (§A.2.1). Nutrients enter the plant primarily through prey capture (Bradshaw and Creelman 1984) and passive uptake of dissolved nutrients in rainwater (Chapin and Pastor 1995). The root system of *S. purpurea* is poorly developed and contributes little to nutrient uptake (Butler and Ellison 2007), although there is some indirect evidence from reciprocal transplant experiments between bogs and fens of more substantial N uptake through roots (Bott et al. 2008).

Plants respond differently to changes in absolute versus relative amounts of nutrients (appendix B), and we emphasized in chapter 2 that strategies to acquire and distribute resources in the context of trade-offs among plant traits were a key feature in the evolution of botanical carnivory. In this chapter, we show that *S. purpurea* responds very differently to inputs of inorganic nutrients from atmospheric deposition versus inputs of organic nutrients from prey capture. We show that nutrients from these two sources affect plant growth, morphology, and nutrient uptake in different ways, only some of which conform to expectations or trait spectra derived from broad syntheses of noncarnivorous plants (Wright et al. 2004; Díaz et al. 2016). Ultimately, these responses scale up, in terms of levels of biological organization, to differences in long-term demography (chapter 6) and, in space, to persistence at the continental scale (chapter 7).

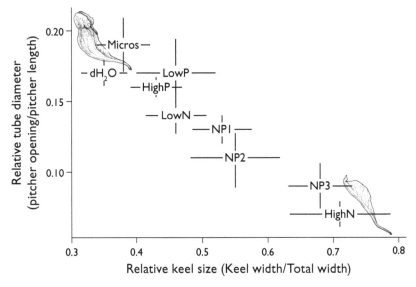

FIGURE 3.1. Relative size of leaves produced by *Sarracenia purpurea* as a function of inorganic nitrogen (as 0.1 [LowN] or 1.0 [HighN] mg NH_4-N/L as NH_4Cl/L), phosphorus (as 0.025 [LowP] or 0.25 [HighP] mg PO_4-P/L as NaH_2PO_4), or micronutrients (10% Hoagland's solution [Micros]) added weekly in one of nine combinations (LowN, LowP, High N, HighP, Micros; three stoichometric manipulations: LowN + LowP [NP1], HighN + HighP [NP2], and HighN + LowP [NP3]; and distilled water controls [dH$_2$O]) to replicate sets of 10 pitchers. Figure redrawn from Ellison and Gotelli (2002).

3.1 STOICHIOMETRIC MANIPULATIONS OF *SARRACENIA*

3.1.1 Effects of Soluble N from Atmospheric Sources

In two separate field studies (Ellison and Gotelli 2002; Wakefield et al. 2005), we manipulated absolute nutrient concentrations and relative nutrient ratios fed to *S. purpurea* and found a variety of shifts in morphology and nutrient stoichiometry. In the first study, Ellison and Gotelli (2002) removed the pitcher liquid and replaced it with different solutions of nutrients in distilled water. Treatments included controls, micronutrient additions, P-only additions, N-only additions, and three treatments that altered N:P ratios with relatively modest changes in absolute concentrations (figure 3.1).

The plant responses were dramatic and occurred within a single growing season. As the N:P ratio increased, leaves developed relatively small pitchers with large

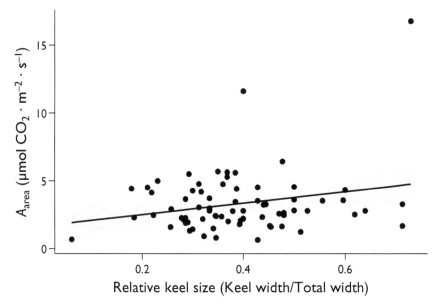

FIGURE 3.2. Maximum photosynthetic rates of pitchers as a function of keel size. Relative keel size increased (from left to right) with increasing N addition (see figure 3.1). Line fit using robust linear regression. Redrawn from Ellison and Gotelli (2002).

flat keels (figure 3.1; keel illustrated in figure A.5). In the extreme, plants produced phyllodia, flat leaves that were all keel and no pitcher. Phyllodia developed as a result of some of the nutrient manipulations, but this leaf type also is occasionally seen in unmanipulated populations (frequency $\sim$5% of all leaves produced, based on >10,000 plants sampled from 77 bogs in the 6 New England states and 22 years of observations at Molly Bog in Vermont) and occurs in response to a variety of environmental stressors (Mandossian 1966).

Phyllodia have a higher density of stomata and are photosynthetically more efficient than normal pitchers (figure 3.2; Ellison and Gotelli 2002). Based on the results from our short-term experimental study, the phenotypically plastic response of the plant to shift from production of pitchers to phyllodia seems to be adaptive: when atmospheric N is relatively scarce (controls), plants form pitchers to catch arthropod prey, but when atmospheric N deposition is high relative to P (N and high N:P additions), plants shifts to "foraging" on carbon (Ellison 2020).

We measured plant morphology and pore-water chemistry in 13 ombrotrophic bogs (herbaceous wetlands that have virtually no contact with groundwater) and poor fens (nutrient-poor herbaceous wetlands that are in contact with groundwater) in Connecticut, Massachusetts, and Vermont. Average pitcher keel width was positively correlated with pore-water N concentrations, which in turn largely reflected estimated atmospheric inputs (figure 3.3; Ellison and Gotelli 2002). The US National Forest Service has adopted this measure of *Sarracenia* leaf morphology, which presumably integrates N and P inputs over several seasons, as an inexpensive biological indicator of atmospheric N deposition (Pardo et al. 2010).

3.1.2 Effects of Nutrient Inputs from Supplemental Prey

In a second field experiment that we conducted in another bog in northern Vermont (Belvidere Bog), we varied the quantity of all food resources but maintained fixed stoichiometric ratios (Wakefield et al. 2005). In this experiment, we supplemented the natural diet of pitcher plants with an additional 2–14 frozen houseflies per week (7 treatment levels plus the controls) over the course of a single growing season. Houseflies have an average N:P:K ratio of 103:8:9 (Wakefield et al. 2005), which is very similar to the stoichiometric ratio of these three macronutrients in ants (12:1.5:1; Farnsworth and Ellison 2008), which are the most common prey captured by *Sarracenia* (Ellison and Gotelli 2009). This experiment altered the absolute amount of nutrients provided but retained a balanced nutrient ratio across all treatments. At the end of the experiment, we measured and harvested the leaves and assayed the nutrient content (N, P, K, Mg, Ca) in the leaves and the pitcher fluid.

In contrast to when we manipulated nutrient ratios from atmospheric inputs (figures 3.1 and 3.2; Ellison and Gotelli 2002), feeding plants additional prey did not alter the size or morphology of leaves by the end of the season, nor did it affect intraseasonal measures of photosynthetic rate (Wakefield et al. 2005). But this does not mean that *Sarracenia* was unresponsive to increases in absolute resource levels. Instead of shifts in plant growth and morphology, nutrient concentrations in the leaves and in the pitcher water were strongly affected.

Wakefield and colleagues' (2005) most striking result was that tissue N concentration almost doubled across the range of treatments, from 1% in the controls to almost 2% in pitchers fed the most prey (figure 3.4; Wakefield et al. 2005); at this high-N tissue concentration, terrestrial plants start to become attractive to insect herbivores (Mattson 1980; R. Denno, *personal communication* to NJG). The enrichment from the experimental treatments was apparent from a distance: the highest prey-addition treatments were literally stuffed with rotting flies and could be smelled in the field from several meters away. The fluid in these pitchers was almost completely anoxic (see also chapter 12), and [NH$_4$] was almost an order

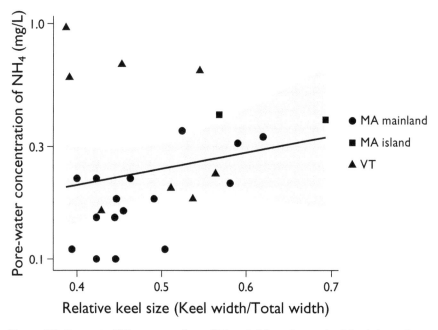

FIGURE 3.3. Pore water NH_4 concentration at 26 bogs in Massachusetts (mainland sites and two bogs sampled on the island of Martha's Vineyard shown in different symbols) and Vermont. Redrawn from Ellison and Gotelli (2002).

of magnitude higher than in the controls (0.4 versus 3.5 mg NH_4-N/L; Wakefield et al. 2005).

We anticipated that this severe enrichment would have led to an asymptote or decline in the rate of N uptake by *Sarracenia* leaves because such high concentrations of N in leaves could have been toxic. Instead, Wakefield et al. (2005) never observed an upper limit for uptake of N by *S. purpurea*. Phosphorus and K uptake also increased with additional feeding. Magnesium levels differed by treatment but not in the same order as the number of flies fed to the plants, while Ca levels were not affected by prey additions (figure 3.4).

However, the uptake rates were not equivalent for different macronutrients, and the stoichiometry of leaf tissues shifted with feeding levels. Specifically, the N:P ratio in the tissues of control plants was just over 17, significantly higher than in any of the prey-addition treatments. Because this ratio also exceeded 15, the data imply that the plants experiencing ambient levels of atmospheric inputs and prey availability were P limited (Koerselman and Meuleman 1996; Aerts and Chapin 2000; Ellison 2006). This conclusion also was

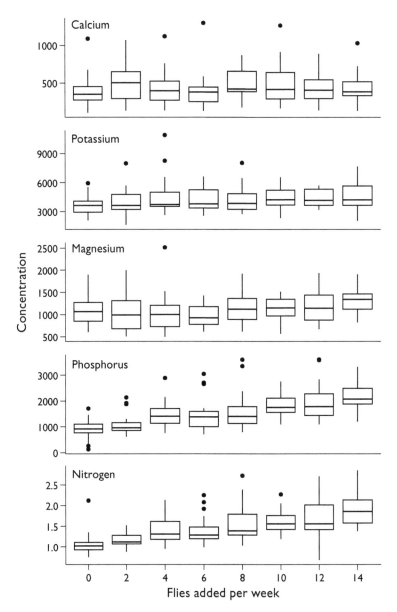

FIGURE 3.4. Nutrient concentrations in pitchers fed varying numbers of flies by Wakefield et al. (2005). Data are for first pitchers only; $n = 25$ plants per treatment. Values are pooled across all five harvests. Units of concentration are in mg/L for all elements but nitrogen, which are in percent.

supported by the near-zero [P] measured in the pitcher fluid of all plants and by the observation that tissue P rose faster at higher treatment levels than did tissue N.

Taken together, these results suggest that mineralized phosphorus was taken up immediately by plant tissues (Ellison 2006). Although both N and P uptake increased with additional feeding, plants consistently took up relatively more P than N, and the plants shifted from being P limited (N:P > 16) to being N limited (N:P < 14). After a sharp drop in the N:P ratio from 17 in the controls to 11.8 in the lowest feeding level (2 flies/week), the ratio then dropped more slowly to 9.1 in the highest feeding treatment (14 flies/week) (figure 3.5; Wakefield et al. 2005).

3.1.3 Synthesis of Supplemental Feeding Experiments

Ternary plots of N, P, and K values for leaf tissues from plants in both experiments indicated that plants were strongly P limited or co-limited by N and P (figure 3.5). Unlike the experiment in which we added inorganic N to pitchers (figures 3.1 and 3.2), we saw no changes in leaf morphology or photosynthetic rates when plants were fed additional prey. As plants shifted from being P limited under ambient conditions to being N, K, or N+K limited following prey enrichment, the plants continued to produce pitchers to capture additional prey, resulting in a more stoichiometrically balanced diet of N, P, and other macro- or micronutrients. In contrast, the "fast-food diet" provided by atmospheric inputs of N alone led to stoichiometric imbalances and a shift toward production of phyllodia.

Both of these experiments were run for only a single season. However, *S. purpurea* stores excess nutrients absorbed within a single season and reallocates them to pitcher and flower production in subsequent seasons (Butler and Ellison 2007). If we had continued our experiments for multiple seasons, we might have seen a positive growth response, as has been observed in prey- or nutrient-enrichment experiments using other carnivorous plant species (Ellison 2006). Other experiments with *S. purpurea* and subsequent demographic modeling similarly revealed positive growth responses following prey enrichment but not when plants were supplied with additional inorganic nutrients (chapters 6 and 7).

3.2 NUTRIENT ADDITIONS IN OTHER PHYTOTELMATA

Trait plasticity and responsiveness to nutrient additions characterize other carnivorous plants (reviewed by Ellison 2006), epiphytes, and bromeliads. For example, *S. purpurea* in bogs and fens of Wisconsin appeared to be N, not P

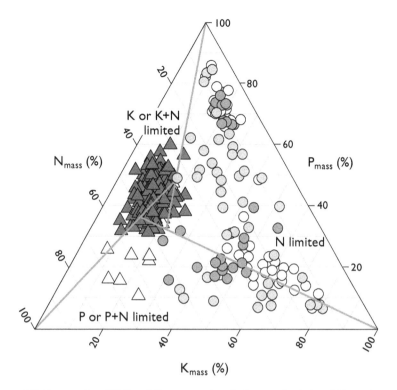

FIGURE 3.5. Ternary plot of N, P, and K concentrations in *Sarracenia* pitchers fed supplemental prey or inorganic nutrients (N_{mass} (%) = $\frac{N}{N+10P+K}$, P_{mass} (%) = $\frac{10P}{N+10P+K}$, K_{mass} (%) = $\frac{K}{N+10P+K}$). Triangles = prey addition experiment (Wakefield et al. 2005); circles = inorganic nutrient addition experiment (Ellison and Gotelli 2002). Symbol shading increases with relative N content of the different experimental treatments. White symbols: no supplemental prey or distilled water or only micronutrients; pale gray symbols: high P, low P, low N; increasingly dark gray symbols: N:P = 1:1.25 (NP1 in figure 3.1), N:P = 4:1 (NP2), N:P = 40:1 (NP3); black symbols: high N or supplemental prey. Nutrient-limitation boundaries (gray lines) are based on the criteria of Olde Venterink et al. (2003); the central triangle is where N, P, and K all are stoichiometrically balanced.

limited, but plants in high-N bogs produced more phyllodia, and plant morphology and leaf [N] converged to those of the local environment when plants were reciprocally transplanted between bog and fen (Bott et al. 2008). N:P ratios of epiphytic *Tillandsia* L. also converged in reciprocal transplant experiments among four sites in the Atacama desert (González et al. 2011). In that habitat, most N uptake was from fog water, which did not differ in its nutrient profile among sites.

Although much of the early work on plant carnivory emphasized N limitation, carnivorous plants grow in habitats in which virtually all macro- and micronutrients are in short supply (Adamec 1997). In such habitats, arthropod prey provide a variety of macro- and micronutrients and trace elements. For example, in addition to uptake of N and P, the pitcher plants *Sarracenia purpurea* ssp. *purpurea* f. *heterophylla* (Eaton) Fern., *Cephalotus follicularis* La Billadiere, and *Heliamphora nutans* Bentham actively take up and translocate K, Mn, and Fe to varying degrees (Adlassnig et al. 2009). Although there are two carnivorous bromeliad species [*Brocchnia reducta* Baker and *Catopsis berteroniana* (Schult. & Schult. f.) Mez; Cross et al. 2018], all other tank bromeliads lack specialized traps for insect prey. Nonetheless, tank bromeliads accumulate large volumes of water and detritus and respond to N and P additions (Ngai and Srivastava 2006; González et al. 2014). The C:N ratio of plant tissues in Neotropical bromeliads largely reflects allochthonous detrital inputs, with some inputs from photosynthetic algae in sunny habitats (Farjalla et al. 2016).

Carnivorous plants and bromeliads are strongly nutrient limited, take in most of their nutrients through leaf surfaces rather than root tips, and are surprisingly plastic in their growth. At large scales, N:P ratios generally increase toward the equator, consistent with the idea that P limits plant growth at low latitudes in more weathered soils, whereas N limitation is more important in younger temperate and high-latitude soils (Reich and Oleksyn 2004; cf. Hou et al. 2020). Clinal gradients in nutrient stoichiometry could mirror spatial gradients in prey availability, especially in rates of atmospheric deposition in different habitats.

For example, chronically high levels of atmospheric N deposition in the northeastern United States may be responsible for stoichiometric P limitation in *S. purpurea* and other wetland plants. Across a set of 20 poor fens in Vermont and Massachusetts, N:P and N:K ratios of *S. purpurea* increased with elevation along a geographical transect from the southeast (near Boston, Massachusetts) to the northwest (in Vermont; Gotelli et al. 2008). The chemistry of pore-water samples from these bogs mostly reflected atmospheric deposition and varied spatially. From the southeast to the northwest, concentrations of dissolved organic carbon (DOC), dissolved organic nitrogen (DON), Cu, Mg, NO_3, Al, and K all decreased. Although it is difficult to tease apart cause and effect, this spatial variance parallels a gradient of decreasing human population density, urbanization—and probably nutrient enrichment (Ellison and Gotelli 2002).

In contrast, pore-water concentrations of SO_4 and Al were highest in the western sites, and SO_4 concentrations increased with elevation. These patterns may have reflected atmospheric inputs from the Ohio River Valley leading to increased acidic deposition, which is known to cause Al to be leached from soils (Schaberg et al. 2000). Nutrient stoichiometry of *S. purpurea* leaf tissues reflects both the physiological constraints of overall nutrient limitation and the particular nutrient ratios that plants experience in different sites.

Similarly, on an elevational gradient in Puerto Rico from lowland forest to dwarf cloud forest, biomass and N:P ratios of the bromeliads *Guzmania berteroniana* (Schult. f.) Mez, *G. lingulata* (L.) Mez, and *Vriesea sintenisii* (Baker) L.B. Sm. & Pittend. increased with elevation, again probably reflecting greater nutrient inputs from atmospheric sources at higher elevations (Richardson et al. 2000). Controlled nutrient additions in a factorial experiment with N and P demonstrate that *Tillandsia elongata* H.B.K. var. *subimbricata* (Bak.) L. B. Sm. is co-limited by both N and P concentrations, and tissue N:P ratios changed in response to both nutrients (Zotz and Asshoff 2010).

At smaller scales, we would expect to find spatial variation among populations in nutrient content of leaves. For example, at the same site in central Argentina where González et al. (2011) worked, the N:P ratio of leaf tissue in epiphytic *Tillandsia capillaris* Ruiz & Pavón was higher when they grew attached to telephone wires (5.7) than on trees (4.5; Abril and Bucher 2009). The difference probably reflected the greater input of atmospheric N to the plants growing on telephone wires.

Converting nutrient ratios in both plant data and environmental samples to a multidimensional "stoichiometric niche" (González et al. 2017) is a promising framework for scaling up plant responses and growth. We begin exploring that framework in chapter 4. Subsequently, in chapter 7, we discuss continental-scale demographic consequences of spatial gradients in nutrient deposition.

3.3 SUMMARY

Like all carnivorous plants, *S. purpurea* evolved under conditions of chronically low nutrient availability. Since 1850, anthropogenic emissions have greatly increased atmospheric inputs of N. Within only one or two *Sarracenia* lifespans, the ambient environmental conditions in which *S. purpurea* grows has shifted from N limited or N+P co-limited to P limited. This novel environmental background likely has resulted in contrasting responses of *S. purpurea* and other carnivorous plants to prey additions versus additions of soluble, inorganic nutrient solutions. The former is a "balanced diet" that includes needed sources of N, P, K, micronutrients, and cofactors that together can enhance growth. In contrast, atmospheric deposition is a "fast-food diet" of empty calories with excess soluble N. Like the protagonist in the 2004 documentary film *Super Size Me*—whose 30-day fast-food diet led to rapid weight gain and metabolic dysfunction—*S. purpurea* responds to inorganic N additions through dramatic phenotypic plasticity and physiological changes. These acclimatory responses have negative effects in the longer term, because inorganic N additions also increase juvenile mortality. Ultimately, these physiological and demographic shifts can lead to population decline as overall death rates exceed birth rates in environments with high rates of N deposition (chapters 6 and 7).

CHAPTER FOUR

Scaling Up

Stoichiometry, Traits, and the Place of Sarracenia
in Global Spectra of Plant Traits

> *I like a balanced universe.*
> —God, in Atwood (2004: 9)

From a continually expanding dataset of more than 100,000 plant species (Kattge et al. 2020), ecologists and evolutionary biologists have identified predictable scaling relationships among core physiological and morphological traits (Wright et al. 2004; Reich 2014; Díaz et al. 2016). In this chapter, we explore the relationships between traits, stoichiometry, and environmental conditions observed for *Sarracenia purpurea* and other carnivorous plants in the broader context of these global spectra and global stoichiometric relationships (González et al. 2017). Under ambient conditions, the carnivorous plants as a group are in the statistical tails of these trait distributions (Ellison and Farnsworth 2005; Farnsworth and Ellison 2008; Ellison and Adamec 2011). One goal of our research with *S. purpurea* has been to understand why carnivorous plants do not follow all the scaling "rules" and what we can learn about "universal" scaling laws from apparent exceptions to them.

4.1 GLOBAL PLANT TRAIT SPECTRA

Central to macroecology is the identification of scaling relationships—ideally derived from physical constraints or first principles—that link form, function, distribution, and abundance. All plants fix carbon, acquire a finite range of nutrients, and allocate the derived resources (e.g., sugars, starches, nucleic acids, proteins, ATP) to physiological processes responsible for maintenance, growth, survivorship, and reproduction. Scaling relationships among key traits that drive and maintain these processes, including A_{mass}, leaf mass per unit area (LMA in g/m^2), and mass-based concentrations of N and P (N_{mass}, P_{mass} in mg/g), are broadly consistent across tens of thousands plant species (chapter 2, figure 2.1; Wright et al. 2004; Díaz et al. 2016). These predictable scaling relationships have been called the "worldwide leaf economics spectrum" (Wright et al. 2004), the "'fast–slow' plant economics spectrum," and the "global spectrum of plant form and function" (Díaz et al. 2016).

4.1.1 Traits

The key process underlying these global spectra is the return—expressed as some measure of fitness—on the investment (or allocation) of photosynthate, acquired nutrients, and water (Wright et al. 2004; Reich 2014). For most plants, sugars and nutrients are the essential resources used to construct roots, leaves, and stems. Roots, in turn, return water and nutrients; leaves return more photosynthate; and stems support the leaves and are anchored by the roots. Thus, the worldwide leaf economics spectrum focuses on A_{mass}, LMA, N_{mass}, and P_{mass}, along with dark respiration rate and leaf longevity (Wright et al. 2004). By also including water as an essential resource, Reich (2014) extended the scaling relationships to include roots and stems and identified scaling relationships in traits that link roots and stems with leaves. Reich (2014) further posited that the combination of physical constraints and strong selection for traits and life histories would lead to all organs and physiological or resource allocation processes of individual taxa falling on a "fast–slow" spectrum. That is, if a plant has a high (low) rate of resource acquisition, it would also have fast (slow) rate of growth and rapid (slow) physiological processes. Díaz et al. (2016) brought these two ideas together in an analysis of six related functional traits: adult plant height, stem specific density (mg/mm^3), leaf area (mm^2), LMA, N_{mass}, and diaspore mass (mg).

4.1.2 Trait Data

The first global compilation of plant trait data was the Global Plant Trait Network (GLOPNET) dataset (Wright et al. 2004), which included traits from 2548 species representing 219 families and 175 sites. Only three carnivorous plant species—*S. purpurea* (two records) and two *Drosera* species—were included in the GLOPNET dataset. A much larger database of plant traits is the TRY Plant Trait Database (Kattge et al. 2020). The version of TRY from which we extracted trait data for comparison with *S. purpurea* and other carnivorous plants (version 4.1; February 4 2018), included nearly 7 million records of 1800 traits from nearly 150,000 plant taxa (the original GLOPNET dataset is now incorporated into TRY). Whereas Wright et al. (2004) focused on scaling relationships of species-specific mean trait values, Kattge et al. (2020) reported that intraspecific differences in traits accounted for up to 40% of the overall variation. Carnivorous plants are much better represented in TRY (338 species) and appear in approximately the same proportion as they are found among vascular plants as a whole (0.2%; Ellison and Adamec 2011). However, for most of these carnivorous plant species, data on only a few traits are available in TRY.

For the analyses in this chapter, we focused principally on nutrient traits (mass-based N, P, and K) and stoichiometric ratios because these are the most plentiful trait data for carnivorous plants. We also synthesized and discuss the more limited data on photosynthesis and construction costs of carnivorous plants to further identify their positions along global trait spectra. Specifically, we aggregated data from the following TRY datasets: GLOPNET (dataset 20; Wright et al. 2004), BiodiERsA-PEATBOG campaign: plant carbon, nitrogen, and phosphorus (CNP) (dataset 324; Bjorn Robroek, *unpublished*); Global Leaf Element Composition Database (dataset 81; Watanabe et al. 2007); and the Americas N & P database (dataset 129; Kerkhoff et al. 2006).

These data were merged and duplicate entries were removed. We then added additional carnivorous plant data from our own work (Ellison and Gotelli 2002; Ellison et al. 2004; Ellison and Farnsworth 2005; Wakefield et al. 2005; Ellison 2006; Farnsworth and Ellison 2008; Gotelli et al. 2008; Ellison and Adamec 2011; the Harvard Forest Data Archive datasets HF109, HF112, HF146, HF168; and the Carnivorous Plant dataset TRY File Archive 72).

4.2 CARNIVOROUS PLANTS IN GLOBAL TRAIT SPECTRA

4.2.1 Nutrient Concentrations

Carnivorous plants normally have leaf N_{mass} and P_{mass} values that are much lower than those of non-carnivorous plants (Ellison 2006) and below absolute levels at which each nutrient limits physiological processes and growth (20 mg/g of N, 1 mg/g of P) (figures 4.1–4.3; Koerselman and Meuleman 1996; Aerts and Chapin 2000). Leaf K_{mass} is similarly low in terrestrial carnivorous plants but higher in aquatic carnivorous plants (Ellison 2006; Ellison and Adamec 2011). However, in neither group do measured tissue concentrations of K reach absolute limiting levels (8 mg/g; Olde Venterink et al. 2002, figures 4.2 and 4.3).

The median leaf N_{mass} for field or lab-grown carnivorous plants that have not been fed supplemental prey or inorganic nutrients is 12.6 mg/g, well below the 10th percentile for forbs, deciduous shrubs, and deciduous trees, and below the 25th percentile for graminoids. Similarly, the median leaf P_{mass} for carnivorous plants (0.94 mg/g) is well below the 10th percentile for forbs, deciduous shrubs, and deciduous trees but modestly above that of graminoids. Median K_{mass} of carnivorous plants ranges from 7 mg/g (terrestrial species) to 16 mg/g (aquatic species); both values fall within expected ranges of non-carnivorous species (Ellison 2006; Ellison and Adamec 2011).

Overall, despite their generally herbaceous habit, macronutrient concentrations of carnivorous plants are more similar to those seen in evergreen shrubs and trees

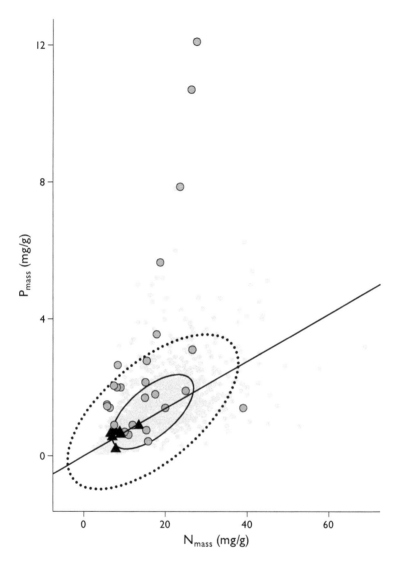

FIGURE 4.1. N_{mass} and P_{mass} (mg/g) of non-carnivorous plants and carnivorous plants. Each gray point represents the average N_{mass} and P_{mass} of a non-carnivorous plant species (TRY data, $n = 1324$ species); solid and dashed ellipses are the bivariate 50th and 95th percentile confidence bounds for the TRY data. Carnivorous plants include *Sarracenia* species (black triangles) and other carnivorous plant species (dark gray circles). The reference line is the N:P ratio (14.5), where neither N nor P is stoichiometrically limiting.

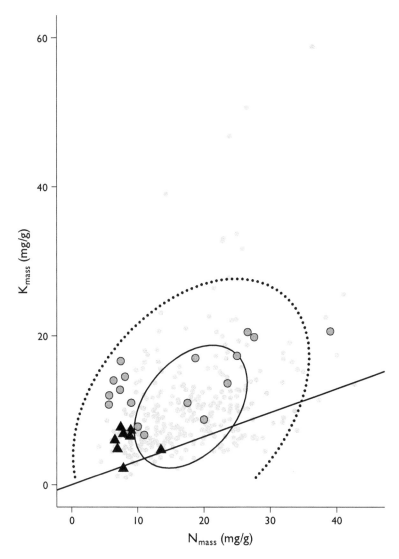

FIGURE 4.2. N_{mass} and K_{mass} (mg/g) of non-carnivorous plants and carnivorous plants. Each gray point represents the average N_{mass} and K_{mass} in a non-carnivorous plant species (TRY data, $n = 1324$ species); solid and dashed ellipses are the bivariate 50th and 95th percentile confidence bounds for the TRY data. Carnivorous plants include *Sarracenia* species (black triangles) and other carnivorous plant species (dark gray circles). The reference line is the N:K ratio (3.1), where neither N nor K is stoichiometrically limiting.

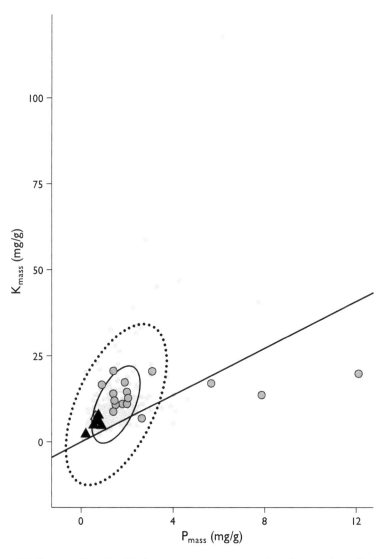

FIGURE 4.3. P_{mass} and K_{mass} (mg/g) of non-carnivorous plants and carnivorous plants. Each gray point represents the average K_{mass} and P_{mass} in a non-carnivorous plant species (TRY data, n = 1324 species); solid and dashed ellipses are the bivariate 50th and 95th percentile confidence bounds for the TRY data. Carnivorous plants include *Sarracenia* species (black triangles) and other carnivorous plant species (dark gray circles). The reference line is the P:K ratio (3.4), where neither P nor K is stoichiometrically limiting.

than they are to forbs or graminoids (Ellison 2006). Although evergreen trees and shrubs have some of the lowest measured values of photosynthetic nutrient use efficiency (PNUE: Wright et al. 2005), they are still photosynthetically more efficient in their nutrient use than are carnivorous plants (Ellison 2006; Ellison and Adamec 2011).

4.2.2 Nutrient Stoichiometry

Even when their tissue nutrient concentrations are above a critical minimum, *Sarracenia* species and other carnivorous plants often are stoichiometrically P-limited (figures 4.1–4.3). Plotting all three nutrients together on a ternary plot, as suggested by Olde Venterink et al. (2002), reveals that the majority of carnivorous plants—including *Sarracenia* species—fall within the stoichiometric space of non-carnivorous ones, but those that do not are strongly P or P+N limited (figure 4.4).

As we illustrated in chapter 3, the relationship between local environmental conditions and the morphological expression of pitcher shape (i.e., keel width or "phyllode-ness") by *S. purpurea* is sensitive to variation in N:P inputs (figure 3.1). For carnivorous plants, the local environment is generally nutrient poor, bright, and waterlogged (Givnish et al. 1984, 2018); light and water are rarely limiting, but nutrients are. In their evolutionary past, carnivorous plants relied on the capture of individual insect prey and occasional insect swarms as their primary source of nutrients. In today's world, nutrient subsidies can also come from atmospheric inputs, surface runoff, or even experimental feeding by ecologists.

4.2.3 Stoichiometric Effects of Supplemental Prey on Carnivorous Plants

Although there have been many studies that have explored the effects of prey addition on carnivorous plant growth (Ellison 2006), only a handful have measured tissue nutrient concentrations. *Sarracenia purpurea* fed supplemental prey preferentially absorbed P while absorbing N and K in proportion to their concentration in the flies (Wakefield et al. 2005). Thus, ratios of N:P and K:P in pitcher-plant tissues shifted toward much higher absolute P concentrations and consequent stoichiometric N or K limitation (figures 4.5 and 4.6). Parallel increases in N_{mass} and K_{mass} did not change N:K ratios, however (figure 4.7). Plotting these results on a ternary plot revealed that preferential uptake of P led plants to move into the 50th percentile of N:P:K space occupied by noncarnivorous plants (figure 4.8).

Farnsworth and Ellison (2008) fed different amounts of ground wasps (*Dolichovespula maculata* [L.]) to 10 species of greenhouse-grown *Sarracenia*. Ground wasps have elemental ratios (C:N = 5.99:1; N:P:K = 10.7:1.75:1.01)

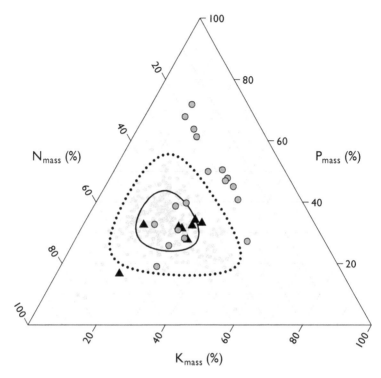

FIGURE 4.4. Ternary plot illustrating potential stoichiometric limitation of N, P, or K in carnivorous plant genera. Light gray circles are N_{mass}, P_{mass}, and K_{mass} for $n = 283$ non-carnivorous plants from TRY; black triangles are N_{mass}, P_{mass}, and K_{mass} of *Sarracenia* species, and dark gray circles are N_{mass}, P_{mass}, and K_{mass} values for other carnivorous species). Solid and dashed ellipses are the bivariate 50th and 95th percentile confidence bounds for the TRY data.

similar to those of the ants that make up the preponderance of the prey captured by *S. purpurea* (chapter 3, §3.1.2; Farnsworth and Ellison 2008). Like Wakefield et al. (2005), Farnsworth and Ellison (2008) observed a decrease in tissue N:P of fed plants (figure 4.5) and a rapid translocation of P from older to younger leaves. Although Farnsworth and Ellison (2008) measured K_{mass}, they did not examine N:K or P:K ratios. Our reanalysis of those data revealed patterns similar to those reported by Wakefield et al. (2005) (figures 4.6 and 4.7). Through preferential uptake of P from prey, leaf N_{mass} and P_{mass} shifted more toward the group centroid of non-carnivorous plants (figure 4.8). Similar results were obtained by Chapin and Pastor (1995) for *S. purpurea* and Chandler and Anderson (1976) for *Drosera whittakeri* Planch.

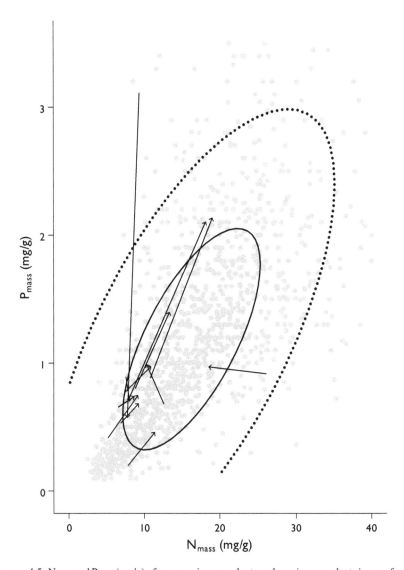

FIGURE 4.5. N_{mass} and P_{mass} (mg/g) of non-carnivorous plants and carnivorous plants in prey feeding experiments. Gray points represent average N_{mass} and P_{mass} of non-carnivorous plants (TRY data, $n = 1324$) and are bounded by bivariate 50% (solid) and 95% (dotted) confidence ellipses. Arrows represent results of prey addition experiments (Wakefield et al. 2005; Farnsworth and Ellison 2008; Chapin and Pastor 1995; Chandler and Anderson 1976). Arrow bases are average nutrient concentrations in control plants and arrow tips are average nutrient concentrations in fed plants.

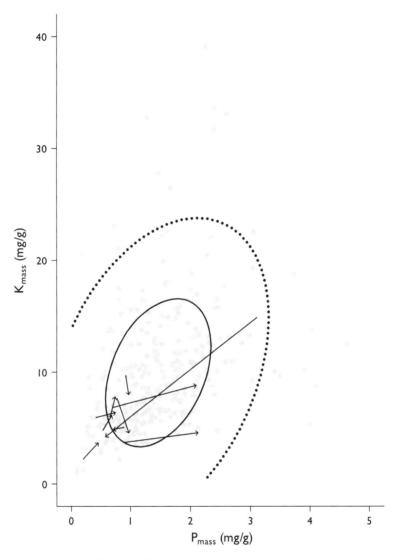

FIGURE 4.6. P_{mass} and K_{mass} (mg/g) of non-carnivorous plants and carnivorous plants in prey feeding experiments. Gray points represent average P_{mass} and K_{mass} of non-carnivorous plants (TRY data, $n = 291$) and are bounded by bivariate 50% (solid) and 95% (dotted) confidence ellipses. Arrows represent results of prey addition experiments (Wakefield et al. 2005; Farnsworth and Ellison 2008). Arrow bases are average nutrient concentrations in control plants and arrow tips are average nutrient concentrations in fed plants.

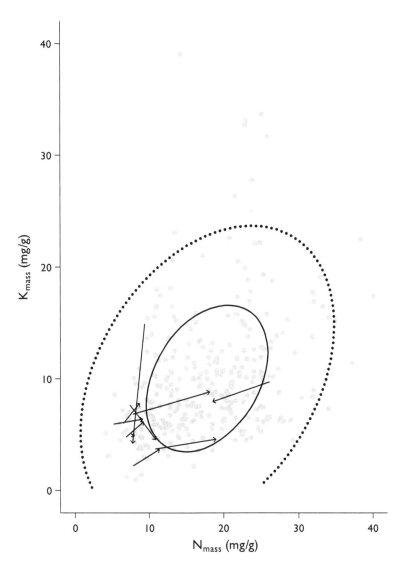

FIGURE 4.7. N_{mass} and K_{mass} (mg/g) of non-carnivorous plants and carnivorous plants in prey feeding experiments. Gray points represent average N_{mass} and K_{mass} of non-carnivorous plants (TRY data, $n = 305$) and are bounded by bivariate 50% (solid) and 95% (dotted) confidence ellipses. Arrows represent results of prey addition experiments (Wakefield et al. 2005; Farnsworth and Ellison 2008). Arrow bases are average nutrient concentrations in control plants and arrow tips are average nutrient concentrations in fed plants.

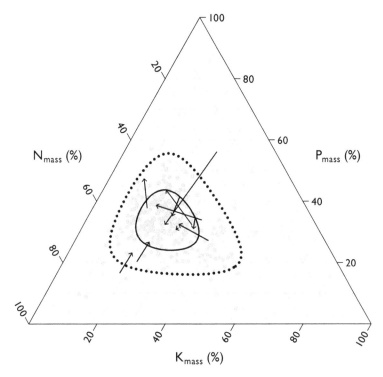

FIGURE 4.8. Ternary plot illustrating stoichiometric effects of prey addition on carnivorous plants. Light gray circles are data for $n = 283$ non-carnivorous plants from TRY; solid and dashed ellipses are the bivariate 50th and 95th percentile confidence bounds for the TRY data. Each arrow represents the results of a single prey addition experiment. The arrow base is the average nutrient concentration in the control plants and the arrow tip is the average nutrient concentration in plants that received supplemental prey. Experimental data from Wakefield et al. (2005) and Farnsworth and Ellison (2008).

4.2.4 Stoichiometric Effects of Adding Inorganic Nutrients to Carnivorous Plants

The prey addition experiments described in the previous section suggested that *Sarracenia* species and at least one species of *Drosera* preferentially take up P from prey. Experiments in which inorganic nutrients, including NH_4^+, NO_3^- and PO_4^+, were fed directly to these species exhibited much greater effects in the same direction (figures 4.9–4.12). When *S. purpurea* was fed inorganic N alone as NH_4Cl, it took up all the supplemental N (Chapin and Pastor 1995; Ellison and Gotelli 2002). *Drosera whittakeri* fed inorganic N as $Ca(NO_3) + (NH_4)_2SO_4$ also took up all the supplemental N (Chandler and Anderson 1976). Similarly, when

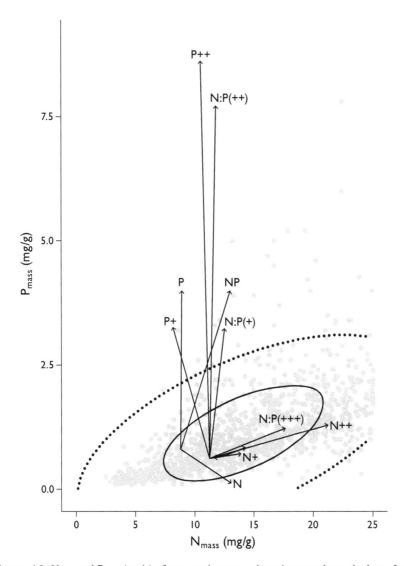

FIGURE 4.9. N_{mass} and P_{mass} (mg/g) of non-carnivorous and carnivorous plants, the latter from three nutrient-addition experiments. Gray points represent average N_{mass} and P_{mass} in non-carnivorous plants (TRY data, $n = 1324$) and are bounded by bivariate 50% (solid) and 95% (dotted) confidence ellipses. Arrows represent results of nutrient addition experiments. Arrow bases are average nutrient concentrations in control plants and arrow tips are average nutrient concentrations in plants that received supplemental nutrients. N, P, NP: nutrient addition treatments to *S. purpurea* from Chapin and Pastor (1995). N+, P+, N++, P++: low(+) and high(++) nutrient additions of N or P to *S. purpurea* from Ellison and Gotelli (2002). N:P(+), N:P(++), N:P(+++): low (+), medium (++), and high (+++) N:P ratio addition treatments to *S. purpurea* from Ellison and Gotelli (2002). Along the N++ vector are two short, unlabeled vectors for N and P additions to *Drosera whittakeri* from Chandler and Anderson (1976).

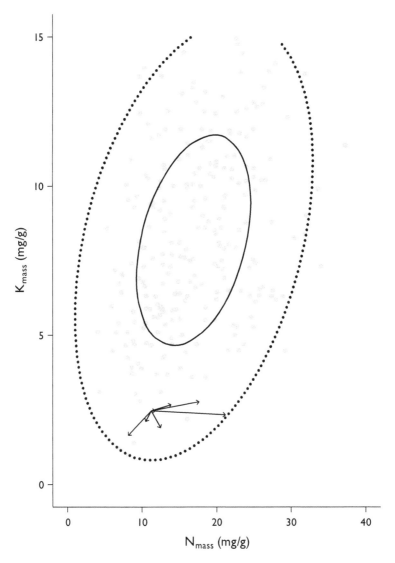

FIGURE 4.10. N_{mass} and K_{mass} (mg/g) of non-carnivorous plants and carnivorous plants, the latter from Ellison and Gotelli (2002). Gray points represent average N_{mass} and K_{mass} in non-carnivorous plants (TRY data, $n = 305$) and are bounded by bivariate 50% (solid) and 95% (dotted) confidence ellipses. Arrows represent results of the nutrient addition experiment. Arrow bases are average nutrient concentrations in control plants and arrow tips are average nutrient concentrations in plants that received supplemental nutrients.

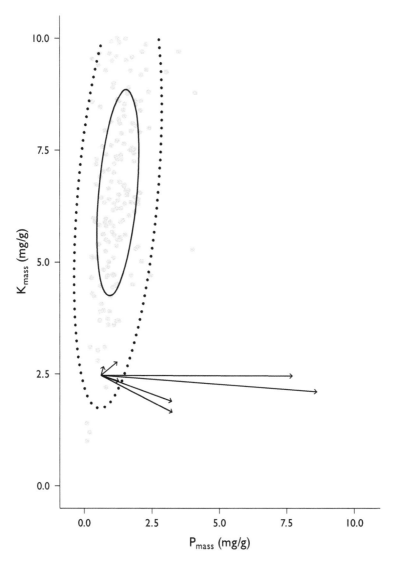

FIGURE 4.11. P_{mass} and K_{mass} (mg/g) of non-carnivorous plants and carnivorous plants, the latter from Ellison and Gotelli (2002). Gray points represent average P_{mass} and K_{mass} in non-carnivorous plants (TRY data, $n = 291$) and are bounded by bivariate 50% (solid) and 95% (dotted) confidence ellipses. Arrows represents results of the nutrient addition experiment. Arrow bases are average nutrient concentrations in control plants and arrow tips are average nutrient concentrations in plants that received supplemental nutrients.

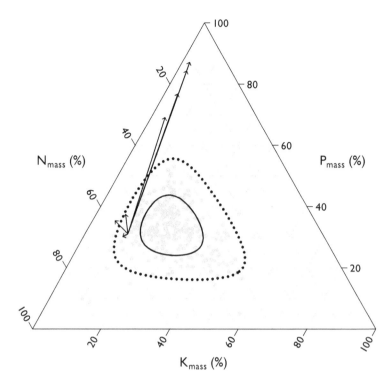

FIGURE 4.12. Ternary plot illustrating stoichiometric effects of inorganic nutrient addition on *S. purpurea*. Light gray circles are data for $n = 283$ non-carnivorous plants from TRY; solid and dashed ellipses are the bivariate 50th and 95th percentile confidence bounds for the TRY data. Each arrow represents the results of adding inorganic nutrients to *S. purpurea* (data from Ellison and Gotelli 2002). The arrow base is the average nutrient concentration in the control plants and the arrow tip is the average nutrient concentration in plants that received supplemental inorganic nutrients.

S. purpurea was fed P alone as KH_2PO_4 (Chapin and Pastor 1995) or NaH_2PO_4 (Ellison and Gotelli 2002), it took up all the available P. But when fed mixtures of inorganic N and P (as $NH_4Cl + NaH_2PO_4$) in various ratios, the plants took up virtually all the P but very little of the N (figures 4.9 and 4.11; Ellison and Gotelli 2002).

This result was not unexpected because addition of NH_4 rapidly induces phosphatase production by *S. purpurea* (Gallie and Chang 1997). *Drosera whittakeri* fed KH_2PO_4 responded similarly (Chandler and Anderson 1976). K was measured in all these experiments, but was not manipulated. However, even when K was added as part of the PO_4 salt, the plants preferentially took up the P (figures 4.10 and 4.11). Among carnivorous plants overall, phosphatases are produced more frequently and in greater concentrations than other hydrolytic enzymes, suggesting convergent evolutionary responses to P limitation in carnivorous-plant habitats.

How do these data compare with those for the non-carnivorous plants in the TRY database? In contrast to the prey-addition experiments (figure 4.8), preferential uptake of inorganic P moved *S. purpurea* and *D. whittakeri* well outside even the 95% confidence bounds of nutrient stoichiometry of non-carnivorous plants (figures 4.9–4.12). This pattern illustrated that environmental conditions can lead to substantial intraspecific variability in leaf traits, as has been observed in other taxa (e.g., Zaoli et al. 2017; Fajardo and Siefert 2018; Osnas et al. 2018). It also suggested that available nutrients may play a much greater role in stoichiometric properties of plant tissues than would be expected from global spectra of plant traits. However, it is also important to note that these experiments all were relatively short term, and plants may regulate tissue nutrient stoichiometry through reallocation over longer time scales.

4.2.5 Photosynthesis and Construction Costs

Because carnivorous plants typically have low photosynthetic rates and low PNUE (Ellison 2006), we would expect them to be very conservative in how they allocate photosynthate and nutrients to the construction of new traps. Construction costs for a particular organ (e.g., leaf, trap, stem, root) are measured in units of g glucose/g dry mass, which provides one measure of speed on Reich's (2014) "fast–slow" spectrum. Unscaled construction costs measure the absolute costs and benefits of an organ, but when scaled relative to A_{mass}, they yield an estimate of the marginal costs (Givnish et al. 1984, 2018) or "payback time" of a trap: the amount of time that a leaf must photosynthesize at its average rate to recover (or amortize) the amount of carbon used to construct the trap (Poorter et al. 2006). Because carnivorous plants grow slowly, photosynthesize at low rates, and have a low PNUE, they are at the slow end of the slow–fast plant spectrum. Thus, we predicted that the construction costs of traps (which are modified leaves; appendix A, §A.2.1) would be correspondingly high.

But surprisingly, we found that construction costs of traps, as well as roots and rhizomes, of a wide range of carnivorous plants (species of *Sarracenia, Darlingtonia, Drosera,* and *Nepenthes*) were substantially lower than the construction costs of leaves or roots of non-carnivorous plants (Karagatzides and Ellison 2009). At first glance, this result implied that traps are relatively inexpensive, perhaps because they are built from relatively cheap total structural carbohydrates instead of expensive compounds and structures needed for photosynthesis (Poorter and Villar 1997).

However, when we scaled construction costs to A_{mass}, the payback time for constructing a trap turned out to be very long, ranging from ≈ 500 hours/g dry mass of a *S. rubra* trap to > 1500 hours/g dry mass of a *S. purpurea* trap (Karagatzides and Ellison 2009). If we assume ≈ 10 hours/day of sun available for photosynthesis, this payback time translates to ≈ 50–150 days of photosynthesis to create a single trap (Karagatzides and Ellison 2009). This payback time is nearly an order of magnitude longer than that estimated for leaf construction of other herbaceous perennials (Williams et al. 1989) or adult trees (Poorter et al. 2006).

The long payback times of *Sarracenia* traps may be ameliorated somewhat by the long lifespan of the plants and their traps; efficient retention, resorption, and recycling of nutrients within and between traps (Butler and Ellison 2007); seasonal variability in A_{mass}; and production of photosynthetically more efficient flat phyllodia (Ellison and Gotelli 2002; Pavlovič et al. 2007). However, first-year leaves are more photosynthetically active and trap insects more efficiently than older leaves (Fish and Hall 1978), so it may be difficult for the plant to ever recover its construction costs, particularly in high-latitude populations where the growing season each year can be <90 days. In short, carnivorous plants seem to be among the slowest of the slow plants on the slow–fast spectrum.

4.3 SYNTHESIS

Recent compilations of plant traits measured for more than 100,000 species have revealed predictable scaling relationships among core physiological and morphological traits. Based on their rates of resource acquisition, individual growth, and physiological processes, most plant species fit on a spectrum from "fast" to "slow," with strong positive correlations among these three rates. Compared to non-carnivorous plants, *Sarracenia* is near the slow end of the spectrum, with low photosynthetic rates and growth rates, and a relatively high cost of generating new traps and flower buds. For stoichiometric traits, some aquatic carnivorous plants have leaf P_{mass} that are substantially higher than most other plants. However, in the genus *Sarracenia*, tissue nutrient concentrations of N_{mass}, P_{mass}, and K_{mass} are comparable to most other terrestrial plants.

As a group, carnivorous plants show consistent responses to nutrient additions, but the patterns are sharply different for inorganic nutrients versus nutrients derived from prey. When carnivorous plants are supplemented with prey, they simultaneously take up several nutrients, and the net trajectory of nutrient uptake keeps the plants within the ranges observed for non-carnivorous species. However, when carnivorous plants are supplemented with inorganic nutrients, tissue nutrient ratios depart greatly from the data for non-carnivorous plants. Both N and

P in plant tissues are highly elevated in response to enrichment from inorganic nutrients.

However, these results are derived from short-term nutrient addition experiments, and it is unlikely that these highly elevated tissue concentrations can be sustained in the long run. The traits and stoichiometry of carnivorous plants reflect adaptations to environments with low nutrients, few competitors, and bright sunlight. It is unclear how *Sarracenia* will respond to the dramatic increases in atmospheric N deposition over the past century, but several lines of evidence suggest that the plants are now P limited because of excess N inputs from atmospheric sources.

Part II
Demography, Global Change, and Species Distribution Models

CHAPTER FIVE

Context

Demography, Global Change, and the Changing Distributions of Species

The greatest shortcoming of the human race is our inability
to understand the exponential function.
—Albert A. Bartlett (1969 et seq.)

In Part II, we tackle the problem of how atmospheric nutrient inputs affect the local population growth of *Sarracenia purpurea* and its long-term persistence at regional and continental scales. This work blends two major themes of ecological research: predicting local population dynamics using demographic models and forecasting occurrences of species at regional-to-continental scales using species distribution (a.k.a. environmental niche) models.

In this chapter, we frame our approaches to demographic and species distribution modeling and discuss why we think the two need to be integrated. In chapter 6, we first summarize our short-term, small-scale experiments from which we estimated the parameters of a stage-structured demographic model of *S. purpurea* (appendix C provides a refresher on the details of demographic models). These experiments were informed by our earlier work on the physiological and morphological responses of *S. purpurea* to nutrient additions and anthropogenic stressors (chapters 3 and 4). We then convert our deterministic, stage-structured model into a stochastic one to provide quantitative forecasts of extinction risk with different long-term scenarios of atmospheric nutrient deposition. These models move well beyond simple equilibrium thinking and explore year-to-year projections of population growth as a function of nutrient inputs that fluctuate because of variations in climate and political winds.

In chapter 7, we take a spatial leap to scale up demographic processes from two small New England bogs to the entire North American continent. At this scale, we use current and projected continent-wide measures of atmospheric nutrient deposition as inputs into the demographic models we developed in chapter 6. We use this analysis to forecast where, within its current range, *S. purpurea* populations will thrive and expand or decline to extinction. We compare these predictions with those from conventional species distribution models (SDMs), which use only information on where populations are presently located in geographical and climatic space to forecast future suitable habitats (appendix D). Even when they are fitted with the same set of environmental predictor variables, demographic models and SDMs give radically different forecasts for the persistence of *S. purpurea*. We argue from this example that ecologists and conservation biologists should incorporate demographic principles and data into models forecasting species distributions in changing environments.

5.1 BACKGROUND

Demography—the analysis of births and deaths in a population—holds a central place in ecology and evolutionary biology (Caswell 2006). Core concepts and theories, such as density dependence (Turchin and Taylor 1992), *r* and *K* selection (Pianka 1970), the theory of island biogeography (MacArthur and Wilson 1963), and the theory of evolution by natural (or sexual) selection (Endler 1986), cannot be tested without data on rates of birth, population growth, and death. Demographic processes are usually studied at local scales within single populations or local metapopulations; incorporating demography into models and forecasts of larger-scale patterns and processes, such as those derived from species distribution models, is difficult and uncommon.

Like demography, geographical patterns of species occurrences and the mechanisms underlying them are also central concerns of ecology and evolutionary biology. For example, descriptions of biomes, ecoregions, and vegetation types depend on accurate occurrence records of plant species (e.g., Holdridge 1947; MacArthur 1972; Olson et al. 2001). Tests of community "assembly rules" (chapter 9) and macroecological associations likewise depend on predictable occurrences and co-occurrence records of species and their relationship to environmental variables (e.g., Diamond 1975; Shade et al. 2018). Because species distributions are described at regional or continental scales, demographic processes occurring within local populations appear on first glance to be unrelated to modeling or forecasting species distributions.

5.2 SDMS, DEMOGRAPHY, AND ANTHROPOGENIC DRIVERS: MOVING BEYOND TEMPERATURE

Both demographic modeling and species distribution modeling have taken on new urgency in recent decades. Even conservative climate models predict an increase of 2–5 °C over the next century (IPCC 2014). In response to these forecasts—and to growing public awareness of the reality of climatic change (Luis et al. 2018)—

many ecologists have been forecasting whether local populations or entire species are declining to extinction (e.g., Sinervo et al. 2010) and how species' geographical ranges will change in the near future (Lavergne et al. 2010; Sequeira et al. 2018). There are other possible outcomes than extinction (Walther et al. 2002), including migration (Higgins and Richardson 1999), range expansion into previously unoccupied areas (Kelly and Goulden 2008), and persistence accompanied by plastic responses in phenotype (Charmantier et al. 2008) or by evolutionary adaptation (Hoffman and Sgrò 2011). If we can successfully forecast these range shifts, we also may be able to forecast the species composition of local assemblages (Botkin et al. 2007) as they form so-called "novel ecosystems" (Hobbs et al. 2009; Collier and Devitt 2016; Evers et al. 2018).

In this part of the book, we emphasize three issues that have been neglected in the development and application of SDMs. The first is that, for many species, temperature may not be the most important factor that will cause their ranges to shift over the next century (e.g., Larsen et al. 2011; Currie and Venne 2016). The second issue, which follows directly from the first, is that some species may be very sensitive to other environmental drivers that change in parallel with, or in opposition to, climatic warming (Vitousek 1994; Vaughan and Gotelli 2019). Finally, demographic processes interact with individual responses to the environment to determine whether a population can establish and persist in its current range or in a new location (Keith et al. 2008; Fois et al. 2018; Ureta et al. 2018). Probabilities of species occurrences ultimately depend on the sizes of local populations, their short-term fluctuations, and their long-term trajectories (Ehrlén and Morris 2015).

5.2.1 Weak Responses to Temperature

It may seem surprising that temperature is not the primary global-change driver that is changing distributions for all species. We do not dispute that the higher temperatures accelerating the melting of polar ice (Alley et al. 2005) may drive some high-latitude species to extinction and may kill other species restricted to high mountains (Dullinger et al. 2012). It is similarly true that many tropical species already are living near their upper physiological thermal limits (Sunday et al. 2014). With additional warming and no nearby thermal refuges, these warm-climate species also may go extinct, creating a tropical "vacuum" of lost diversity (Colwell et al. 2008).

But different challenges may face temperate-zone animals and plants at lower elevations. If their geographical ranges are large, their populations collectively may already have experienced the warming that is forecast to occur over the next century. Even within a single population, individuals annually experience very large temperature ranges between winters and summers. Although the current global rate of climatic change may be unprecedented, the average temperatures that will be reached in the near future already are experienced routinely by many organisms on a seasonal or even daily basis.

Sarracenia purpurea is a good case in point. Its geographical range extends from the Canadian Rocky Mountains to the Florida panhandle (appendix A, figure A.1). In parts of the Canadian Rockies, the annual temperature range—from the hottest summer day to the coldest winter night—exceeds 50 °C. At many sites throughout the range of *S. purpurea*, daily temperature changes exceed 20 °C. Like any plant, *S. purpurea* cannot escape high temperatures by moving, but it can acclimate physiologically (up to a point) and could tolerate major climatic changes in situ. At the community level, the indirect effects of climatic warming could bring greater pressures from competitors, herbivores, diseases, or predators. These indirect effects may prove to be as important for *S. purpurea* population dynamics as the direct effects of increased temperature and changes in the length of the growing season (see also Diamond et al. 2016).

5.2.2 Nutrient Enrichment as Another Global-Change Driver

Burning fossil fuels directly contributes to atmospheric warming while also releasing NO_x (Jaeglé et al. 2005), and the use of fertilizers (both synthetic and natural) releases NH_4 (Krupa 2003). Whereas climatic change has occurred over millions of years, the 1.5-fold increase in the biosphere of biologically reactive nitrogen has occurred over less than a century (Townsend and Howarth 2010) and has few paleoclimatic analogs (Junium et al. 2018).

For a long-lived organism like *S. purpurea*, the time during which atmospheric deposition of N compounds has increased precipitously amounts to <10 generations of exposure, which is extremely short in evolutionary terms (Lallensack 2018; cf. Reznick et al. 2019). As the human population continues to grow beyond its current size of nearly eight billion people, the increased demand for additional food is projected to lead to an increase in agriculture and fossil-fuel consumption. Under a "business as usual" scenario, global N use is projected to increase from 94 kg/ha in 2005 to 160 kg/ha in 2050 (Tilman et al. 2011). For many plants and animals, the global intensification of N use and the changes in N cycling may be a more immediate threat than the long-term consequences of climatic change. Even though concentrations of N (or other pollutants) do not appear as a column in climatic databases such as WorldClim (Fick and Hijmans 2017) or PRISM (Prism Climate Group 2004) that serve as inputs into SDMs, these anthropogenic environmental drivers should be incorporated into our models and forecasts. At the same time, understanding the role of N in changing plant distributions is complex. At low concentrations, N compounds act as fertilizers, enhancing plant growth and flowering. At high concentrations, however, N can be toxic, particularly to young individuals.

5.2.3 The Importance of Demographic Effects

Just as they neglect nutrient enrichment and environmental drivers other than temperature and precipitation, cookie-cutter applications of SDMs ignore underlying demographic processes. A key feature of the analysis of demographic models is the consideration of the effects of age, size, or life-history stage on population growth, persistence, and extinction (appendix C). SDMs, however, assume that all individuals in a population are affected equivalently by their local environment and that there is no variation in how individuals—much less life-history stages—might respond to climatic variability or directional change (appendix D). But different life-history stages of plants, animals, and humans are affected differently by climate, climatic change, and extreme events (e.g., Ridd et al. 2013; Filewod and Thomas 2014; Laurel et al. 2017; de Souza et al. 2018; Li and Zhou 2019).

For example, as the climate warms and winter shortens, early-season warming is accelerating first-flowering time in many plant populations (e.g., Davis et al. 2015; Park et al. 2018). At the same time, the increased thermal variability is leading to occasional "late" frosts and "false springs" (Allstadt et al. 2015). Farmers and foresters have long known that late spring frosts associated with temperature warming or elevated CO_2 can ruin a year's crop (e.g., Lutze et al. 1998; McKenney et al. 2014; Meier et al. 2018). Risks to non-crop plants from spring-frost damage also are increasing (e.g., Augspurger 2013), affect different species differently (e.g., Ma et al. 2019), and can be ameliorated by demographic and life-history events such as enhanced productivity by leaves later in the growing season (Malis et al. 2016; Zohner et al. 2019). Among animals, warming can affect reproductive effort and allocation (Smith et al. 2013), demographic transitions (Komoroske et al. 2014), and maturity rates (Lepetz et al. 2009; Kordas and Harley 2016). And for humans, demographic effects of climatic change have been documented through archaeological and historical studies (e.g., An et al. 2005; Manning and Timpson 2014).

5.3 NEXT STEPS

Demography and population dynamics are at the core of ecology and evolutionary biology. Climatic warming, and other rapidly changing environmental drivers individually and interactively can affect organisms differently at different points in their life-histories. Yet species distribution models used to forecast population responses to climatic changes rarely include effects other than temperature or precipitation, much less demographic or ontogenetic parameters. At the same time, standard deterministic formulations of age- and stage-based matrix models for population dynamics can give misleading forecasts because they ignore measurement error, environmental change, and other sources of stochasticity that can affect elements of (st)age transition matrices. In the next chapter, we develop stochastic demographic models of *S. purpurea* that begin to address these forecasting challenges. In chapter 7, we integrate demographic models into continental-scale SDMs to forecast persistence, extinction, and distribution of *S. purpurea* as anthropogenic forcing continues to change their key environmental drivers.

CHAPTER SIX

The Small World

Demography of a Long-Lived Perennial Carnivorous Plant

Il n'est pas certain que tout soit certain.[1]
—Blaise Pascal (1897 [Brunschvicg 387])

In this chapter, we start out by modeling local population dynamics with a simple stage-projection matrix model that forecasts population growth of *Sarracenia purpurea* at Hawley and Molly Bogs (the necessary technical details of demographic matrix modeling are presented in appendix C; see also Gotelli and Ellison 2002, 2006b). Then, we combine the demographic data with results from a two-year nitrogen-enrichment field experiment. The result is an empirical function that lets us create transition matrices for particular N-deposition regimes. Next, we use existing time-series data on N deposition to create a simple forecasting model to generate time series of N availability under different deposition scenarios. Finally, we combine all these elements to estimate the probability of species occurrence in 100 years under a range of N-deposition scenarios. In the next chapter, we scale up this model to the continental United States. The resulting geography of rates of population increase or decrease at each pixel on the landscape—visualized as a heat map—provides an illuminating contrast with the more typical projections derived from a standard MaxEnt species distribution model (SDM) built with *S. purpurea* occurrence records and WorldClim climatic variables.

6.1 DEMOGRAPHIC MODELS OF *SARRACENIA PURPUREA*

6.1.1 A Deterministic, Stage-Based Demographic Model for Sarracenia purpurea

Our simple stage-based matrix population growth model for *S. purpurea* divided the population into four mutually exclusive stages: juveniles (rosette diameter <10 cm), non-flowering adults (rosette diameter ≥10 cm), flowering adults, and recruits (seedlings in their first 1–2 years of growth). In the terms of equation C.1

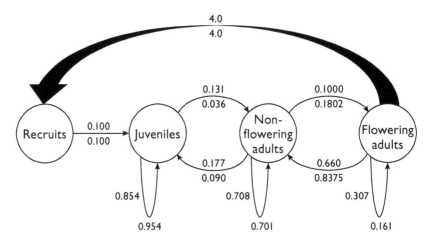

FIGURE 6.1. Loop diagram of the stage-projection model for the demography of *S. purpurea*. Each node represents an exclusive life-history stage, and the arrows represent possible transitions between two stages in each annual time step. The values above and below the arrows represent, respectively, the measured or estimated transition elements for populations monitored at Molly and Hawley Bogs.

(appendix C), n is a 4-element vector containing the number of individuals in each stage in the population at a time step t.

The demography of *S. purpurea* can be illustrated as a loop diagram (figure 6.1), in which the four circular "nodes" represent the four life-history stages and the arrows (directed "edges") specify possible annual transitions between them (the entries a_{ij} of the matrix A in appendix C, equation C.1). An arrow that goes from one stage to itself (a self-loop) indicates the probability that the plant remains in the same stage from one year to the next. There are double-headed arrows between flowering and non-flowering adults because individual adult plants rarely flower in two consecutive years (appendix A, §A.2.3). Double-headed arrows between non-flowering adults and juveniles can represent between-year rosette shrinkage or, less frequently, measurement error. There are only single-headed arrows between flowering adults and recruits, and between recruits and juveniles, because these transitions cannot occur in the reverse direction.

To estimate the transitions for each arrow in figure 6.1, in 1996 we marked cohorts of 100 adult and 100 juvenile plants at Molly and Hawley Bogs. Every year since then, we have revisited the plants and recorded the flowering status, rosette diameter, and several measurements of leaf morphology. The censuses were terminated at Hawley Bog in 2003 after a moose walked through the plot, killing many of the plants and destroying most of the flags and plant markers.

The censuses have continued at Molly Bog to the present day; by 2020, only 3 survivors remained of the original cohort of 200 plants.

We initially used the marked cohorts to estimate average transition probabilities for two consecutive years (1996–1997 and 1997–1998; Gotelli and Ellison 2002). We could not estimate two of the transitions in this model—flowering-to-recruits and recruits-to-juveniles (i.e, seedling establishment)—from the cohort data, so we used other approaches. To estimate the flowering-to-recruit transition, we first used observed data that mature fruits have, on average, ≈1000 seeds each (Ellison 2001; Ellison and Parker 2002). Then, in both bogs, we planted small spatial grids of seeds and observed that on average 4 recruits germinated for every 1000 seeds planted (Gotelli and Ellison 2002). From these data, we estimated a transition element from flowering adults to recruits equal to 4 (i.e., four germinating seedlings per flowering adult).

To estimate the recruit-to-juvenile transition, we "guesstimated" a 90% mortality rate between the recruit and the juvenile stages. Although it is far from ideal to mix data with ad hoc guesses, this patchwork approach combining empirical observations with expert knowledge is fairly common in the construction of stage-based models because not all of an organism's life-history stages can be measured or censused easily or accurately (Crouse et al. 1987; Wiens et al. 2017).

Using the deterministic model (appendix C, equation C.1) with transitions estimated from field experiments and guesstimation, we found that the finite population growth rate was low at both Hawley ($\lambda = 1.00458$) and Molly Bogs ($\lambda = 1.00500$). These are slow exponential growth rates, with corresponding doubling times of 152 and 125 years, respectively (Gotelli and Ellison 2002).

As with many perennial plants (Franco and Silvertown 2004), the most important contributions to population growth rate of *S. purpurea* were the persistence of juveniles (elasticity = 0.43 for Molly Bog, 0.61 for Hawley Bog)[2] and the persistence of non-flowering adults (elasticity = 0.39 for Molly Bog, 0.27 for Hawley Bog). The least important transitions were for recruitment (elasticity = 0.018 for Molly Bog, 0.016 for Hawley Bog) and the survival of recruits to the juvenile stage (elasticity = 0.018 for Molly Bog, 0.016 for Hawley Bog). These results were reassuring because there was great uncertainty in our guesstimation of the recruitment transitions compared to the other, empirically based transitions (Gotelli and Ellison 2002).

Because **A** was constant, the deterministic model forecasted a stable stage distribution that would be observed in an unchanging environment. We were interested in comparing the predicted stable stage distribution from the Hawley and Molly Bog models to the actual stage distribution of plants at Hawley Bog, which were measured using small-scale random plot sampling combined with estimates of total bog area from aerial photographs. Observed differences from

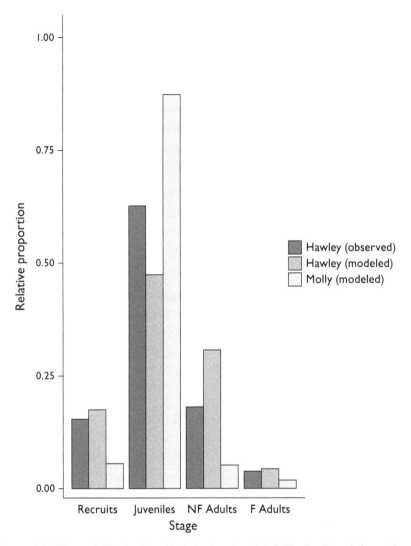

FIGURE 6.2. Observed (Hawley Bog: black bars) and modeled (Hawley Bog: dark gray bars; Molly Bog: light gray bars) proportions of different life-history stages of *S. purpurea*. Projections were derived from the analytic solution to equation C.1 using the parameter values in figure 6.1.

model predictions could result either from insufficient time to achieve a stable stage distribution in the field or because the "environment" (A) is not actually constant (Doak and Morris 1999).

The model predictions did not match the observed stage distribution of plants (figure 6.2). Specifically, the model underestimated the observed proportion of juvenile plants and overestimated the proportion of non-flowering adults. The estimates for the proportion of recruits and the proportion of flowering plants were reasonably close. It is unlikely that this mismatch was because of insufficient time to reach demographic equilibrium, as these bogs have been prominent features of the landscape for thousands of years (Foster et al. 2006; Rayburn et al. 2007) and *S. purpurea* populations likely are equally old. Rather, the differences illustrated in figure 6.2 would be expected for a long-lived perennial plant growing in a variable environment with annual or longer-term environmental fluctuations such as those seen for N-deposition rates in recent decades (chapter 2, figure 2.3). Initial evidence supporting this conclusion was that the fit of the data to the model predictions was better for Hawley Bog (where the data were collected) than for Molly Bog, which has a similar climate but a very different N-deposition regime (Gotelli and Ellison 2002).

To further explore the mismatches between current environmental conditions and the demographic states of their populations, we started with the observed size structure at Hawley Bog, iterated equation C.1, and calculated the stage structure expected at each time step. These trajectories suggested that the populations would reach their equilibrium in about 15–20 years, starting from their observed initial distributions (figure 6.3). Thus, the observed stage distribution now appears to be a snapshot of accumulated changes that the population experienced 15–20 years earlier.

On the one hand, a lag of 15–20 years is moderately short for a plant that can live 50–75 years, and probably corresponds to less than a single generation. The relatively rapid demographic response of *S. purpurea* is consistent with the observation that the plant is phenotypically plastic and very responsive to changes in its pitcher-water chemistry (chapter 3, figures 3.1 and 3.3; Ellison and Gotelli 2002).

On the other hand, a time-step of 15–20 years can represent a substantial change in N-deposition rates in either a negative or positive direction. For example, from 1982–1996, total N deposition at the NADP monitoring station nearest to Hawley Bog (Site MA08) increased by nearly 50% (from 4.7 to 6.8 kg/ha), whereas, for the subsequent 15-year interval, deposition decreased by 47%, to 3.6 kg/ha. In 2017, total N deposition near Hawley Bog was just over half that measured in 1982 (2.6 kg/ha) (chapter 2, figure 2.3). It is unlikely that N deposition will remain constant (see chapter 7, §7.4), and *S. purpurea* is likely to continue to be in demographic disequilibrium.

6.1.2 Stochastic Stage-Based Models

The assumption that A is constant in equation C.1 is rarely true. Projection models based on this assumption can give misleading forecasts in changing climates

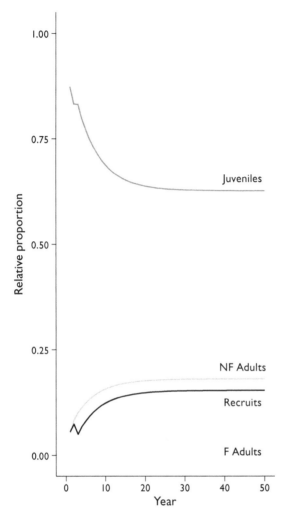

FIGURE 6.3. Modeled changes in the relative proportions of individuals in each population growth stage. The starting proportions were measured in the field at Hawley Bog (black bars in figure 6.2) and the deterministic model (equation C.1) was used to iterate the model at each time step. The transient dynamics disappeared after approximately 15–20 years. NF: nonflowering; F: flowering.

because they ignore two major sources of uncertainty: uncertainty caused by changing environmental conditions and uncertainty caused by measurement error and imprecise estimates of the elements of the transition matrix (Åberg 1992; Keith et al. 2008; Ehrlén and Morris 2015). Our deterministic model (equation C.1) assumed a constant environment and no measurement error in parameter estimates. In contrast, stochastic models can incorporate both environmental change and measurement error. In the remainder of this chapter, we use stochastic models to improve our stage-transition model and demographic forecasts for *S. purpurea*.

Our initial estimates of transition probabilities (figure 6.1) were derived from the standard approach that considered each of them as the proportion of transitions observed for the individuals in each stage in our cohort. We derived better estimates of transition probabilities by sampling from a Dirichlet distribution, which accounts for sample size and the non-independence of the elements within each column of the transition matrix subject to the constraint that the latter must (when mortality is included) sum to 1.0.[3]

We used a statistical bootstrap to capture the uncertainty in estimation, given the cohort numbers we used to generate the data in figure 6.1. Because the recruitment transitions were not based on cohort sampling, we assigned a small sample size ($n = 10$) for these transitions to incorporate their high uncertainty (relative to the other transitions in the table) and we did not incorporate uncertainty into the fecundity estimate. Next, we randomly sampled transition parameters for 1000 transition matrices and calculated the deterministic λ for each one of these (Gotelli and Ellison 2006b).

This analysis put realistic bounds on the uncertainty of deterministic population growth rates. For Molly Bog, 95% of the simulated values of λ fell between 0.98 and 1.04, suggesting that the actual population growth of the Molly Bog population was anywhere from a 2% annual loss to a 4% annual gain, with a probability = 0.42 that the population was decreasing ($\lambda < 1$). For Hawley Bog, the interval was 0.98–1.03, with a probability of decline = 0.40 (figure 6.4; Gotelli and Ellison 2006b). These intervals were larger than we might have hoped for, but they are typical for field demographic studies of this kind (Doak et al. 2005).

Although there was a wide range of uncertainty in these estimates, the deterministic λ values were just above 1.0, yielding an annual rate of increase of only ≈0.5% per year. In this exponential growth model, birth and death rates were closely matched. But a different interpretation could be that the population was regulated, and the low growth rates reflected ongoing density dependence. With density-dependent growth, the deterministic λ may be ≈ 0 because populations will grow rapidly when they are small, but grow slowly or decline when they are large. The net effect would be a measured rate of exponential increase that is very small.

The results of the simulations (figure 6.4) did not suggest density-dependent regulation of these *S. purpurea* populations. Although the estimates of population growth rate were close to zero, they were not unusually close ($p = 0.33$ for Molly Bog and $p = 0.31$ for Hawley Bog). A four-year greenhouse experiment tracking *S. purpurea* from germination through early seedling growth found some evidence for density dependence in survivorship of one- and two-year-old seedlings: nearest-neighbor distances were greater for seedlings that survived than for seedlings that died (figure 6.5). After two years, however, density dependence was no longer apparent, which supported our simulation results. On the other hand, field data for

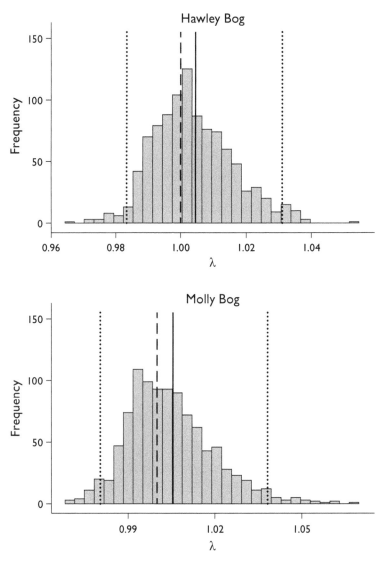

FIGURE 6.4. Estimated uncertainty from measurement error in the population growth rate λ for Hawley Bog (**top**) and Molly Bog (**bottom**). The uncertainty was generated by a Dirichlet sampler applied to the parameters in figure 6.1. Solid line: observed λ; dotted lines: 95% confidence interval on the sampled distribution; dashed black line: $\lambda = 1$, the replacement rate of no growth.

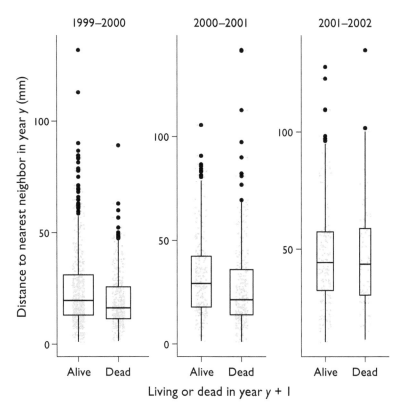

FIGURE 6.5. Density-dependent survivorship in seedlings of *Sarracenia purpurea*. One thousand stratified *S. purpurea* seeds were randomly scattered on a 1 × 2.5-m table filled 10-cm deep with milled *Sphagnum*. Germinants were marked with numbered toothpicks and mapped to the nearest mm. Survivorship was monitored annually from germination in May 1999 through harvest in September 2002. Between 1999 and 2001, surviving seedlings were 20% further away from nearest neighbors than those that died ($P < 0.001$ for 1999–2000; $P = 0.002$ for 2000–2001). After 2001, however, survivorship was independent of distance to nearest neighbor ($P = 0.64$). Widths of the box plots are proportional to the number of seedlings in each year × survivorship category combination. Data from Harvard Forest Dataset HF202.

S. alata in Mississippi suggest some degree of density dependence in its population growth (Brewer 2019).

6.2 EXPERIMENTAL DEMOGRAPHY

Both the deterministic and stochastic models suggested that *S. purpurea* populations were growing slowly, with birth and death rates closely balanced (figure 6.4;

Gotelli and Ellison 2002, 2006b). There was little evidence for density-dependent control of population size, and population size structure lagged estimated current demographic rates by perhaps 15–20 years. These analyses were useful for understanding *S. purpurea* population growth in a constant environment, but they did not give us insight into how current and future environmental change (e.g., changes in N-deposition rates) could affect its population growth. Fortunately, the biology of *S. purpurea* makes it an excellent system for field manipulations that can mimic changes in atmospheric chemistry. Applying these manipulations to cohorts of plants let us examine the demographic consequences of changes in water chemistry from atmospheric deposition.

Methods   In the field, we removed the natural pitcher water and replaced it with liquid in which we precisely controlled the chemical profile. We then plugged the entrance to the pitcher with spun glass wool, which discouraged adult insects from ovipositing and suppressed the development of an animal food web to further control the nutrient inputs to the plant. We applied nine different experimental treatments in which we manipulated nitrogen, phosphorus, micronutrients, and the N:P ratio, and applied these nutrient cocktails to small cohorts of juvenile and adult plants (Ellison and Gotelli 2002). In this "life-table response experiment" (LTRE), we measured the survival, growth, and flowering of these plants over two years and estimated stage transitions using the basic matrix model that we developed for the cohort demography study described in the previous section. We then used parametric bootstrapping to incorporate uncertainty and compare λ values among the different treatments (Gotelli and Ellison 2002, 2006b).

Results   Different nutrient environments had different effects on population growth rate (figure 6.6). Positive log(λ) values were observed for the control, micronutrient, and P-addition treatments. But low, medium, and high N additions led to increasingly negative projections for log(λ), as did a separate array of treatments in which the N:P ratio was systematically increased. Collectively, these results suggested that population growth was probably P limited but not N limited, consistent with our previous analyses of nutrient stoichiometry of leaf tissues (chapters 2–4). The results also suggested that future increases in the absolute concentration of N or the N:P ratio that the plant receives from precipitation could push populations into decline. This projection also was consistent with our observations that current demographic rates of fecundity, growth, and survivorship are in fairly close balance, so additional atmospheric enrichment could push these populations from positive into negative growth (Gotelli and Ellison 2002).

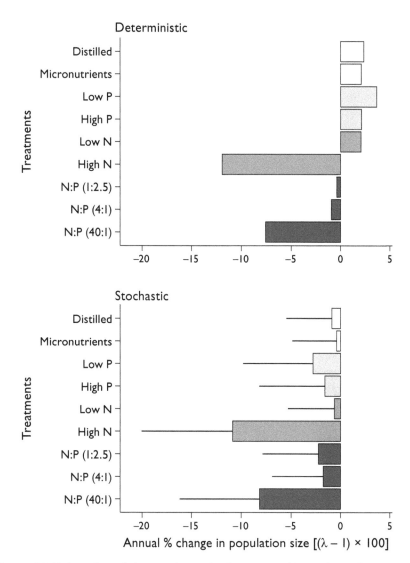

FIGURE 6.6. Estimated population growth rates for *S. purpurea* with experimentally controlled pitcher-water nutrient concentrations. The *x*-axis units are the percentage increase or decrease in population size estimated by the exponential growth model. The results of the stochastic model (lower panel) assume variability represents measurement error. Error bars (shown in only one direction) represent one standard deviation.

6.3 DEMOGRAPHY IN A CHANGING WORLD

We further improved our stochastic demographic model by using our field observational and experimental data in a mechanistic model of a stage-structured population ($\boldsymbol{n}$) that is affected by an environmental driver (E). The model was built on three general equations:

$$E_{t+1} = f(E_t) \tag{6.1a}$$

$$\boldsymbol{A}_t = g(E_t) \tag{6.1b}$$

$$\boldsymbol{n}_{t+1} = \boldsymbol{A}_t \boldsymbol{n}_t \tag{6.1c}$$

The first of these equations (6.1a) describes the state of the environmental driver E in each time step as a stochastic function (f) of its state in the previous time step. The second equation (6.1b) describes a deterministic linking function (g) that generates a stage-transition matrix $\boldsymbol{A}_t$ for each time step t, based on the level of the environmental variable E at time t (E_t). The third equation (6.1c) iterates population growth using the series of transition matrices that were constructed to represent the levels of the environmental driver through time. Thus, equation 6.1 is a non-stationary model in which the transition matrices change at every time step (Gotelli and Ellison 2006b). There is no equilibrium solution, and the fate of the population depends uniquely on its starting size, stage structure, and the particular sequence of environmental changes it experiences. Both the linking function (equation 6.1b) and the population growth equation (equation 6.1c) are deterministic. We allowed for variability through different iterations of equation (6.1a).

6.3.1 Forecasting Nitrogen Deposition

We also needed a model for forecasting future changes in an environmental driver E. Very complex models could have been constructed here, but we preferred a simple iterative time-series model in which the level of E in each year year is a function of the level in only the previous year. For *S. purpurea*, we used N-deposition records from the two NADP stations closest to Hawley and Molly Bogs (MA-08 and VT-99; chapter 2, figure 2.3). We fit first-order autoregressive time-series models to the total atmospheric deposition of nitrogen (from both NH_4 and NO_3). Note that to avoid confusing amount of N deposition with population size (N), we continue to use the variable E to represent this environmental driver.

The best-fitting models for E at the two sites had a similar autoregressive-with-lag-1 ("AR-1") form:

$$E_{t+1} = a + bE_t + \epsilon \tag{6.2}$$

In this model, a is the intercept, representing current N-deposition levels; b is the multiplier for a constant rate-of-change of N deposition; and $\epsilon \sim \mathcal{N}(0, 1)$.

There are two advantages to this kind of model. First, the coefficient b represents the λ for exponential growth in N deposition. This parameter can be used by conservation biologists and land managers to incorporate different long-term deposition scenarios, such as a 1% annual increase ($b = 1.01$) or a 0.5% annual decrease ($b = 0.995$) in N deposition. To forecast extinction risk with a scenario of no change in average deposition, b can be set equal to 1.0.

The second advantage to this model is that it builds in temporal autocorrelation: N deposition in year $y + 1$ is at least partly dependent on N deposition in year y. But because there is a stochastic error term (ϵ), N deposition will tend to drift up or down for extended periods, even though its long-term average will trace a trajectory of exponential increase or decrease. If $b = 1$, the N-deposition trajectory follows a pure random walk, with no long-term deterministic trends. This kind of model is more realistic for a phenomenological forecast of future trends than a pure white-noise model in which the values of each time step are completely independent.

Figure 6.7 illustrates four different N-deposition trajectories projected for 100 years, each with no trend in the average deposition rate ($b = 1.0$). Two of the trajectories used the fitted AR-1 model for Hawley Bog and two used the fitted AR-1 model for Molly Bog. The stochastic variation in these trajectories ultimately was what generated the probability distribution for extinction. And jumping ahead just a bit, figure 6.8 shows the non-stationary population-growth trajectories that were generated by these four N-deposition time series. All the population trajectories were declining, but the autocorrelation in N values generated some sustained intervals of a decade or more with steady increases in population sizes.

6.3.2 Linking N-Deposition Rates to Stage-Transition Matrices

We used data from our nutrient addition experiment to create a realistic link function (equation 6.1b). This function generated a unique transition matrix associated with any particular level of N deposition. In figure 6.6, we showed the estimated λ (with measurement uncertainty) for each of the nine experimental treatments. Here we focus just on the control, low-N, and high-N treatments because these were the ones that could be related to the long-term N-deposition data.

Hindsight did reveal weaknesses in our experimental design. For the purposes of a standard LTRE, the original design was effective for comparing among nine different nutrient-addition treatments and how they affected λ. But when we went on to build the link function, we had data with which to estimate probabilities for only three points over a large and continuous range of potential N values.

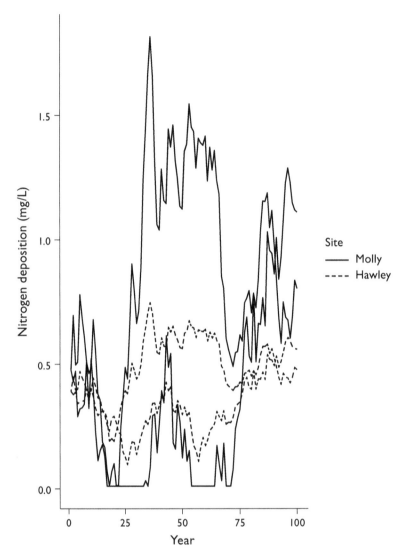

FIGURE 6.7. Autocorrelated time-series trajectories of annual N deposition simulated for Molly Bog (solid lines) and Hawley Bog (dotted lines). The two trajectories at each site represent different underlying random-number series for the error term (ϵ) in the time-series model. At the end of each simulation run, negative values were reset to 0.01 to reflect low N-deposition values. The larger standard deviation of ϵ for Molly Bog generated much more variability between runs than those simulated for Hawley Bog.

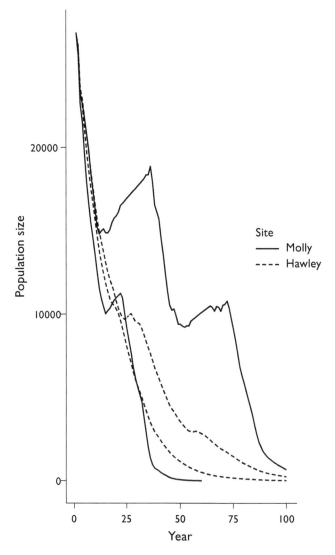

FIGURE 6.8. Four simulated trajectories of total population size of *S. purpurea* populations at Molly Bog (solid lines) and Hawley Bog (dotted lines), Each trajectory was built from the simulated trajectory of N deposition (figure 6.7) combined with the transition matrix function generator (figure 6.9).

Our original experiment had 10 plants each assigned to control, low-N, and high-N additions (Gotelli and Ellison 2002). This worked just fine for an analysis of variance, but for modeling we would have been better off if we had used a regression design (Cottingham et al. 2005) with 30 evenly spaced N levels to identify continuous (and likely non-linear) relationships between transition probabilities and N. Such a design would have yielded ample data for a logistic regression model, which then would have predicted directly the probability of growth or mortality for any level of N addition.

So what could we do with only three data points? Our solution was to plot [N] on a logarithmic scale, connect the three points with straight lines, and make two assumptions about what would happen to transition probabilities for N levels that were higher or lower than our treatments. At the low end, we assumed that transition probabilities would not change if N concentrations were even lower than measured in our control treatment (which began with distilled water that had low but still detectable levels of N in the field). At the high end, we assumed that an additional 10-fold increase in N concentration would lead to 100% mortality for plants exposed to that much N. This was somewhat unsatisfying, because the steepest changes in the function curves were drawn for the extrapolated concentrations for which we do not have any actual data. Nevertheless, this function gave us what we wanted, which was a stage-transition matrix that was tailored for any measured N concentration (figure 6.9).

Note that figure 6.9 has $4 \times 4 = 16$ panels because the function $g(\cdot)$ comprises 16 individual nitrogen functions, one for each of the elements in the transition matrix. Seven of the 16 functions are constants (equal to 0) because they represent transitions in the model (figure 6.1) that do not occur in this life history and could not be affected by N deposition. This set of 16 functions takes the $\log([N])$ at each time t as input and generates a unique stage-transition matrix for each time step of the population-growth model (equation 6.1).

6.3.3 Modeling Population Growth

By combining the stochastic simulation model for forecasting N deposition (equation 6.2) with the deterministic look-up table for building a transition matrix (figure 6.9), we iterated population-growth trajectories (equation 6.1c). Because this was a non-stationary matrix model, we could not solve it analytically for the familiar demographic parameters of population growth rate, stable stage distribution, or elasticity of parameter transitions. Rather, in this kind of model, transient effects are important and the initial population size and stage structure had large effects on times to extinction.

Matrix transition elements

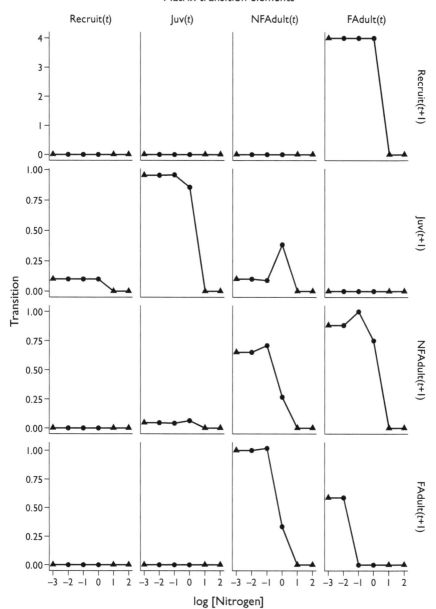

FIGURE 6.9. Look-up functions ($\equiv g(\cdot)$ in equation 6.1b) for effects of log[N] deposition on stage transitions in a non-stationary demographic model. Each panel corresponds to a different stage transition in the life history of *S. purpurea* (figure 6.1). The values indicated by solid circles were derived from experimental manipulations corresponding to the control, low-N, and high-N addition treatments in figure 6.6. The lines and solid triangles represent, respectively, linear interpolations and extrapolations beyond the experimental data.

Figure 6.8 shows some of the complexity in population-growth trajectories that arose from temporally autocorrelated N-deposition rates. When $b > 1.0$, most of these trajectories exhibited a simple pattern of steep exponential decline leading to an inevitable extinction. Even if the population size was still large, rapid extinction occurred if there was a sudden spike in N deposition to a lethal level. At lower [N], some of the populations showed interesting trajectories with quasi-cycles of increase and decrease superimposed over long-term patterns of population decline.

These periodic increases should give conservation biologists pause. Most conservation programs that simply monitor populations and lack data from interventions or experimental work would conclude that a 10-year run of consecutive increases in population size would bode well for long-term persistence. But the model trajectories in figure 6.8 illustrate that these increases are no guarantee of longer-term persistence.

Despite realistic parameter estimates derived from field experiments, our model was still somewhat narrow because it considered only demographic changes that ultimately reflected the physiological responses of individual plants to changing N-deposition regimes. But there are other possible scenarios. For example, increasing N can lead to the loss of *Sphagnum* (Berendse et al. 2001). As nutrient availability increases, species that can outcompete *S. purpurea* may increase in abundance (Bubier et al. 2007).

Other more immediate anthropogenic threats could come from the mining and flooding of peat bogs or the activities of overzealous carnivorous plant collectors and florists, all of which deplete populations and damage fragile bog habitats. The center of species richness for *Sarracenia* is in the southeastern United States, where most of the species grow in sandy seepage swamps and outwash plains (appendix A, §A.1.2). Although atmospheric N deposition is not currently a threat in that part of the United States, >95% of *Sarracenia* habitat has been destroyed there, primarily from the combined onslaughts of urban development, draining of wetlands for forestry, and leaching of pesticides from agriculture into nearby wetlands (Jennings and Rohr 2011; Jennings et al. 2012). Northeastern populations are in better shape because *Sphagnum* bogs often are protected because they are uncommon habitat patches. But even there, the more subtle effects of chronic long-term N deposition may be just as lethal as obvious land-use changes.

To illustrate this, we recorded the percentage of simulated populations that were still present 100 years in the future, regardless of their population size or growth trend. Under current ("no change") conditions of N deposition, these analyses suggested a substantial risk of extinction at Molly Bog ($P = 0.46$) and a relatively small risk of extinction at Hawley Bog ($P = 0.16$). However, the Hawley Bog population was much more sensitive to future increases in N deposition, and the extinction risk over the next century rose steeply with increasing annual deposition rates (figure 6.10).

But these analyses considered only whether the populations persisted for 100 years and did not take into account trends in total abundance over the century. To forecast long-term temporal trends, we estimated λ for each population by fitting a regression of log(*total population size*) versus time for the period in which the model predicted the population was present. The slope of that regression was a simple estimate of r for a model of exponential increase or decrease. We converted this estimate of r back to λ for plotting.

These analyses revealed that even with optimistic annual decreases in N-deposition rates of 5% or more, the expected trends in population growth of *S. purpurea* under most scenarios were still negative (figure 6.11). An upper ceiling of small positive growth rates was imposed because birth and death rates were nearly equal for asymptotic conditions of low N deposition (figure 6.9). Stochastic trends in N deposition therefore caused steep declines in population growth (figure 6.8) but also did not contribute much to positive growth. Between sites, the larger noise term in the time-series models for Molly and Hawley Bogs generated much greater uncertainty in the trajectories of *S. purpurea* population growth at Molly Bog (figures 6.7 and 6.8). But at both sites, population decline was slow and the populations were still relatively large. On the other hand, even under scenarios of strong increase in annual deposition rates, average lifespan of populations was in excess of 150 years (Gotelli and Ellison 2002).

6.3.4 The Future Is Now: Nitrogen Deposition and Extinction Risk in 2020

Our demographic studies on *S. purpurea* were initiated in 1996, and we published quantitative forecasts of extinction risk in Gotelli and Ellison (2002, 2006b). Our earlier estimate of extinction risk for the Molly Bog population was 0.46 (§6.3.3). Since 1996, our initial cohort of 200 tagged plants at Molly Bog has steadily decreased from mortality and tag losses, and by August 2020, only 3 individuals remained from the original cohort.

Four qualitative observations from our continuing work at Molly Bog suggest that its *S. purpurea* population may be declining toward local extinction. First, the ≈20-m-wide ecotone separating the open bog mat from the black spruce/larch forest has expanded over the past 24 years. Larch [*Larix laracina* (Du Roi) K. Koch] and black spruce [*Picea mariana* (Mill.) Britton, Sterns & Poggenburg] recruits are growing up, gradually shading the bog mat and locally adding leaf litter and coarse woody debris. As a result, the open *Sphagnum* mat where *S. purpurea* thrives is shrinking rapidly.

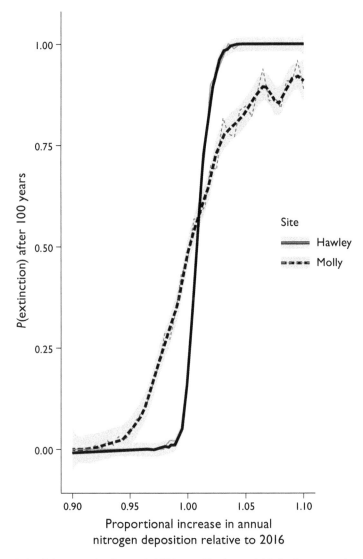

FIGURE 6.10. 100-year extinction probabilities at Hawley and Molly Bogs as a function of changing N deposition. Values on the x-axis are annual proportional changes relative to 2016 in N deposition (from a 10% decrease to a 10% increase). Each datum is the proportion of 100 simulated populations at each N-deposition level that went extinct at some time during a 100-year simulation (figure 6.8).

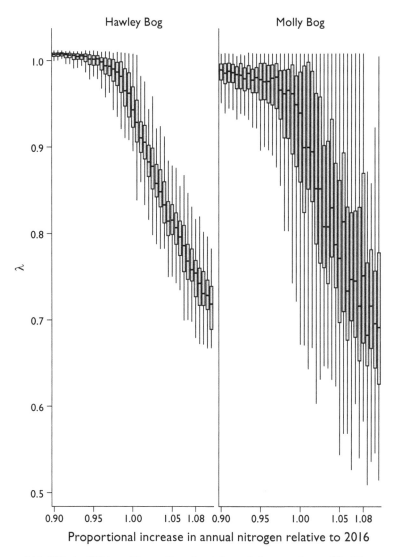

FIGURE 6.11. Effects of N deposition on the estimated population growth rate (λ) of *S. purpurea* at Hawley and Molly Bogs. Values on the *x*-axis are annual proportional changes relative to 2016 in N deposition (10% decrease–10% increase). λ (*y*-axis) is the measured population growth rate for each of 100 replicate population runs for a change in N deposition. The box plots span 50% of the population runs; vertical lines encompass 1.5 times this interval in each direction, and outliers are not shown.

Second, density of *S. purpurea* has decreased noticeably since 1996. Although we have not measured plant density regularly, we estimate it has declined by at least 50% since our research began at Molly Bog. Some of the experiments we ran in the late 1990s and early 2000s would be difficult to replicate now because there are too few spatially separated plants of any size. Concurrently, there has been an even more dramatic decrease in flowering. In the mid-1990s, Molly Bog was carpeted every spring with hundreds of flowering stalks that could not be accurately counted. Now, it is rare to find more than a dozen flowering plants across the entire bog mat. This result is consistent with our finding that low concentrations of N addition increase flowering, but high concentrations decrease it (figure 6.9).

Finally, *S. purpurea* at Molly Bog is less sensitive to N additions. Twenty years ago, we reported that additions of anthropogenic nitrogen transformed leaves from a cup-shaped pitcher to a flattened phyllode (chapter 3, figure 3.1; Ellison and Gotelli 2002). In 2018, Lindsey Pett, a PhD student in Gotelli's lab, repeated parts of experiments of Ellison and Gotelli (2002) as part of her dissertation work on nutrient stoichiometry and regulation of the entire *Sarracenia* microecosystem. Although Lindsey's experimental treatments were not identical to those in Ellison and Gotelli (2002), she did manipulate N and P to alter both absolute nutrient concentrations and N:P ratios. She used an identical protocol for measuring the plants in a single field season but did not detect the morphological changes that Ellison and Gotelli (2002) had reported. The plants no longer exhibit the same phenotypic plasticity that they did in our earlier experiments.

Although we have confidence in our field observations of these qualitative trends, we lack quantitative data on them. We therefore cannot definitively attribute observed changes to chronic nitrogen deposition, increases in warming, or other unmeasured environmental changes that have occurred over the past 20 years. Nevertheless, our initial modeling forecasts of a high risk of local extinction sadly seem to be playing out.

6.4 SUMMARY

We used a three-pronged approach to demographic modeling that incorporated observational data of monitored local populations, experimental data from manipulative life-table response experiments, and models of long-term changes in environmental drivers from published monitoring records. Combining these three elements in a demographic model provided results that could be used for more realistic forecasts of extinction risk given different scenarios of long-term environmental change. Although we applied this approach to changing patterns of N deposition, the same framework can be applied to other potential environmental drivers, such as temperature or precipitation. The detailed analysis of this chapter is based on many years of data collection and experimentation, but it is only representative of two pixels in the landscape of the eastern United States.

CHAPTER SEVEN

Scaling Up

Incorporating Demography and Extinction Risk into Species Distribution Models

*Through the animal and vegetable kingdoms, nature has scat-
tered the seeds of life abroad with the most profuse and lib-
eral hand. . . . Necessity, that imperious all pervading law of
nature, restrains them within the prescribed bounds.*
—Thomas Robert Malthus (1798: 14–15)

In this chapter, we bring together demographic models of *Sarracenia purpurea*
(chapter 6) and our understanding of how it responds to anthropogenic nutrient
deposition (chapter 3) with continental-scale species distribution modeling. Our
goals here are twofold. First, we expand the spatial scale of our demographic
forecasting model (chapter 6) from Hawley and Molly Bogs to the much larger,
patchy landscape of bogs and seepage swamps within the range of *S. purpurea* in
the United States. Second, we expand a basic MaxEnt (Phillips et al. 2006; Elith and
Leathwick 2009) species distribution model (SDM) of *S. purpurea* in continental
North America to include demographic processes mediated by an environmental
driver other than temperature.[1]

7.1 AVAILABLE DATA

Our geographical scaling of demographic forecasting and species distribution
models required spatially extensive data on the distribution of *S. purpurea* and its
associated environmental drivers.

7.1.1 Sarracenia purpurea *Occurrence Data*

Sarracenia purpurea has an enormous range in North America (appendix A,
figure A.1), and the combination of its aesthetic appeal as a carnivorous plant
and its restricted habitat requirements has resulted in an excellent dataset of
its occurrences. We gathered species occurrence records (presence-only) of *S.
purpurea* in the United States from three sources: the survey of *S. purpurea* by

Buckley et al. (2003), data from 78 New England bogs that we sampled from 2006 to 2010, and a compilation of Global Biodiversity Information Facility (GBIF)[2] records from 653 sites that was a subset of data used in a comprehensive MaxEnt analysis of nearly 300 carnivorous plant species by Fitzpatrick and Ellison (2018).

GBIF data are notorious for potential sampling biases and also errors of identification or attribution (Troia and McManamay 2016; Troudet et al. 2017). We did not scrub the data set closely for our analyses, but we did eliminate 11 of the 653 GBIF records for ten sites in Washington and one site in California. These are well outside the recorded natural range of *S. purpurea* and likely indicate plants in cultivation or deliberate introductions. After eliminating these records, our total dataset included 727 unique occurrences in the continental United States with the exception of Alaska.

7.1.2 Environmental and Climatic Data

These 727 unique occurrence records were combined with annual N-deposition data for 2016 from the National Atmospheric Deposition Program (NADP).[3] The NADP data were provided as `.tiff` image files containing smoothed deposition data for the continental United States. We converted these images to spatial data frames in R, projected them into latitude–longitude coordinates (NAD83 geodetic datum), and extracted the N-deposition values for the georeferenced locations of *S. purpurea*. We then converted the N-deposition data from the kg/ha units provided by NADP to mg/L (formulas in Gronberg et al. 2014) for compatability with the units we used in our field experiments. The highest N-deposition sites (in 2016) were concentrated in the upper Midwest and the Ohio River Valley (figure 7.1).

For climatic variables, we used mean annual temperature and mean total annual precipitation from WorldClim (Fick and Hijmans 2017). Terrestrial plant occurrences usually are well correlated with these simple measures of temperature and water availability (Hawkins et al. 2003). A plot of these two climatic variables associated with the occurrences of *S. purpurea* yields a snapshot of its environmental "niche" (figure 7.2).

7.2 CONTINENTAL SCALING OF DEMOGRAPHIC MODELS

7.2.1 Challenges and Simplifying Assumptions

It was not easy to scale up from our demographic studies at two bogs in Massachusetts and Vermont to the set of 727 US sites where *S. purpurea* has been recorded. We did have detailed information on the demographic effects of N

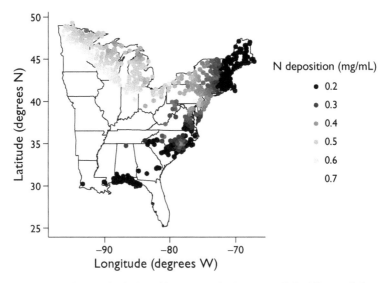

FIGURE 7.1. Annual atmospheric deposition (mg/L) in the eastern United States of nitrogen in 2016. The deposition includes from both NH_4 and NO_3. Each point is a site in which there is an occurrence record of *S. purpurea* ($n = 727$). Note that N-deposition rates are highest in the upper Midwest and the Ohio River Valley and are lowest in southern and coastal regions of the eastern United States.

deposition on *S. purpurea* at two sites (chapter 6). To apply this across its range, however, we needed to assume that the functions that specified the effect of N on each of the 16 demographic transitions in the matrix model (the $g(\cdot)$ functions; chapter 6, equation 6.1b and figure 6.9) operated identically at all sites. This did not mean that all populations would respond uniformly, because the sites varied considerably in their current and past histories of N deposition. Nevertheless, this assumption still ignored the potential for spatial variation in genotype frequencies or in phenotypic responses to N deposition in different parts of the geographical range of *S. purpurea*.

A second issue was that we could not simply apply our original demographic model to each *S. purpurea* population at all 727 sites. That model required counts (or estimates) of the number of individuals of each of the four life-history stages (recruit, juvenile, non-flowering adult, flowering adult). This information was not available for most *S. purpurea* populations. Moreover, our original model was non-stationary: we simulated population trajectories tracking a moving target of N-deposition values and estimated the probability of extinction within a specified time frame. Again, this kind of detailed model was not possible for each site. Instead, we simplified the analysis and used a stationary model to generate long-term projections of population persistence in a constant environment.

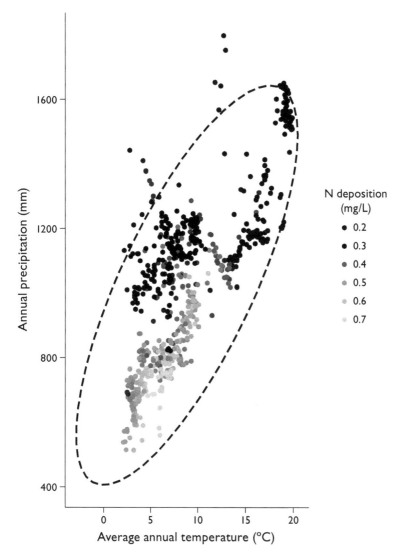

FIGURE 7.2. The environmental niche (with 95% confidence ellipse) of *Sarracenia purpurea* expressed as average total precipitation (1970–2000) versus average annual temperature (°C). Each point represents a location where *S. purpurea* has been recorded in the United States (*n* = 727 locations); gray shading represents the 2016 annual nitrogen deposition (mg/L).

In contrast to the non-stationary models developed in chapter 6, this simplification did not require an estimate of initial population size and stage structure to generate predictions. This type of simplification generally is useful for working with species with many occurrence records but for which local population data are lacking.

7.2.2 Including P Introduced Additional Complexity

Although we had to simplify our model somewhat by assuming stationarity and identical $g(\cdot)$ functions, we did add additional complexity by incorporating effects not only of N but also of P. In the model presented in chapter 6, we considered only temporal variation in atmospheric N deposition, which is monitored annually by NADP and exhibits clear geographical and annual trends in deposition intensity. Here we extend this model to create a response surface in which each transition probability was determined jointly by N and P concentrations.

First, we assigned N concentrations to each of the 727 *S. purpurea* sites based on the NADP N-deposition data for 2016 (figure 7.1). We then estimated values of P deposition at each site from the compilation of Tipping et al. (2014). Using their observations of atmospheric P deposition (mg/L of PO_4-P) at three locations in the continental United States, we obtained the maximum likelihood estimates of parameters for a gamma distribution. We then simulated random draws from this distribution to assign a P concentration to each site.

In our model of demographic effects of N deposition, we used the results from three experimental treatments (control, low-N, and high-N), to create a deterministic look-up function for each of the non-zero transitions solely as a function of N deposition (chapter 6, figure 6.9). To include the effects of P, we incorporated data from the five additional experimental treatments (figure 6.6): low-P, high-P, low N:P ratio, medium N:P ratio, high N:P ratio. Thus, for each transition probability, we had eight experimental data points placed on the two-dimensional N and P plot (figure 7.3).

For each transition probability, we fit the following least-squares regression model:

$$P(\text{transition}_i) = a\text{N} + b\text{P} + c\text{NP} \qquad (7.1)$$

In theory, a logistic regression model could have been fit to the raw experimental data. But in practice, the model parameters frequently will not converge with such sparse data. For this reason, we stuck with an ordinary least squares model fitted to the estimated transition probability for each experimental treatment.

Once all 16 transition probabilities were generated for a given level of N and P, we checked to make sure that the column totals in the transition matrix were not greater than the maximum possible value of 1.0, which would occur if all size classes had 100% survivorship. If the column total exceeded 1.0, we clipped it randomly by 0–2% and distributed the mortality randomly among the different stages. We then estimated λ for this final transition matrix.

Because of variation in simulated P-deposition values and clipping of excessive column totals, the simulated λ values varied for a particular combination of N deposition (determined from the 2016 continental map of N deposition) and P deposition (determined from a single random draw from a gamma distribution). Thus, for each of the 727 *S. purpurea* populations, we ran 100 model simulations to estimate λ. For a set of 100 modeled values, we estimated the long-term probability of population persistence as the proportion of λ values ≥ 1.0. This measure captured uncertainty due to P deposition levels but did not require initial estimates of population size or age structure, so it could be applied to any population for which N and P deposition could be measured or estimated.

This long-term probability of persistence was of interest because we could compare it with MaxEnt's habitat suitability index, which is widely (but incorrectly) interpreted the same way (Yackulic et al. 2013, and see appendix D, §D.2). Before making that comparison, we examined the derived response surface of average λ values across a range of N and P values (figure 7.3).

The poorest performance (darkest shades) was in the lower right corner, which was a combination of high N and low P deposition. However, there was a band of values giving better performance that extended from the lower left (low N and P deposition) to the upper right (high N and P deposition) of the response surface. This band suggested co-limitation of N and P (see also chapter 3, figure 3.4), with the highest possible population growth occurring at very high levels of N and P.

However, this scenario is unlikely to be achieved for two reasons. First, the [P] needed to achieve this level was close to 0.30 mg/L, which exceeded our maximum experimental concentration of 0.25 mg/L by 20%. In contrast, the maximum recorded continental level of atmospherically deposited P was two orders of magnitude lower (0.005 mg/L). Thus, atmospheric sources of P alone would not be sufficient to achieve maximal population growth. In theory, P levels could be boosted if plants acquired additional P from insect prey. Second, however, our previous experiments had shown that when *S. purpurea* receives excessive nitrogen, it shrinks the size of the water-filled pitcher that is used to trap insects (figure 3.1). We conclude that increasing N leads to a demographic dead end because it reduces the capture of insect prey, further exacerbating the unbalanced N:P ratio.

Because the atmospheric contribution of P to *S. purpurea* population growth was so slight, it was better to focus on the lower part of this response surface, within the observed limits of continental deposition of N and P (figure 7.4). In this magnified subsection of the parameter space, higher population growth was achieved at intermediate [N] but was expected to decrease sharply as N deposition increased.

Although this parameter space was informative for thinking about how different combinations of N and P could affect *S. purpurea* demography, the realization of

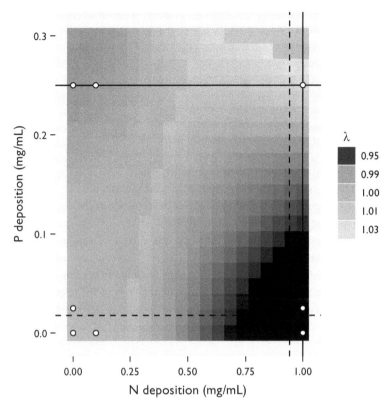

FIGURE 7.3. Response surface for λ, the finite rate of population increase, as a function of N and P deposition rates (mg/L). The eight white circles represent experimental nutrient concentrations that were used to derive the stage-specific effects of nitrogen on stage-transition probabilities (figure 6.9). The solid vertical and horizontal lines represent respectively the maximum [N] and [P] used in the experiment. The dashed vertical line represents the maximum NADP N-deposition rate recorded at the population sites in 2016, and the dashed horizontal line represents the maximum reported P-deposition rate in Tipping et al. (2014). The lighter diagonal swath from lower left to upper right suggests a co-limitation of N and P that affects population growth and persistence.

these scenarios depends on the deposition history at particular sites (figure 7.1). The N and P values for these sites were mapped onto the demographic parameter space of figure 7.3. to generate persistence probabilities at each site.

7.2.3 Continental Forecasts for S. purpurea Persistence

The projection map for *S. purpurea* showed very high rates of extinction in the upper Midwest, through the Ohio River Valley, and in the White Mountains of New

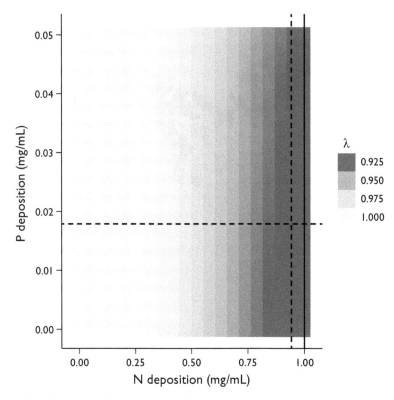

FIGURE 7.4. Response surface for λ, the finite rate of population increase, as a function of N-and P-deposition rates (mg/L). This figure is the subset of the parameter space in figure 7.3 that is restricted to lower levels of P deposition. In this region of the parameter space, the light vertical band indicates an optimal intermediate N-deposition rate that corresponds to the maximum population growth rate. The solid vertical line represents the maximum [N] used in the experiment. The dashed vertical line represents the maximum NADP N-deposition rate recorded at the population sites in 2016, and the dashed horizontal line represents the maximum reported P-deposition rate in Tipping et al. (2014).

Hampshire (figure 7.5). The best prospects for persistence were for populations east of the White Mountains of New Hampshire and the Berkshire Mountains of Massachusetts, populations in the Smokey Mountains, and populations along the coastal plains of the Carolinas and southern Virginia. Populations in the lowest N-deposition areas of Maine and the Florida panhandle had intermediate prospects.

Underlying these prominent geographical trends, the demographic model itself forecast highly non-linear responses to changes in N deposition (figure 7.8). At the lowest N-deposition levels, persistence probabilities were near 50% and decreased slightly as N deposition increased from 0.16 to 0.23 mg/L. Then, between 0.23

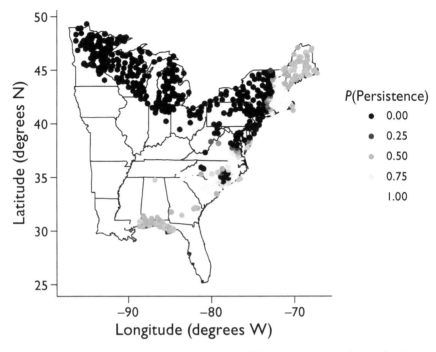

FIGURE 7.5. Demographic predictions of the probabilities of long-term persistence for *S. purpurea* at 727 locations. The predicted probabilities are derived from 100 random simulations of P-deposition values combined with N-deposition values estimated from the site locations in figure 7.1. For each iteration, the N- and P-deposition values specify a population growth rate λ from the parameter space in figure 7.4. The probability of long-term persistence is the proportion of the 100 simulations for which λ > 1.0.

and 0.24 mg/L, persistence probabilities rose very steeply and reached a narrow sweet spot of 1.0 between 0.24 and 0.27 mg/L. Beyond that level, persistence again decreased very steeply and was effectively zero for N-deposition levels > 0.33 mg/L. In contrast, there was little effect of P deposition because on average it does not vary systematically across the continent.

The complex non-linear response to increasing N deposition nevertheless translated into a coherent, spatially structured map of good and bad regions for population persistence (figure 7.5) that reflected transport of pollutants in the upper atmosphere from industrial sources in the upper Midwest and Ohio River Valley. Despite many areas with good prospects for long-term persistence, the overall picture was grim: the demographic model predicted a persistence probability of <50%. Although arrived at with different assumptions, this result corresponded in direction to a slightly negative bioclimatic velocity forecast for 2015–2060 that was estimated independently for *S. purpurea* populations by Fitzpatrick and Ellison (2018).

7.3 FORECASTING THE FUTURE DISTRIBUTION OF
SARRACENIA PURPUREA

7.3.1 A MaxEnt Model for Sarracenia purpurea

We used a standard species distribution model, MaxEnt (Phillips et al. 2006), to forecast the future distribution of *S. purpurea* in the United States. Using the nomenclature of Peterson et al. (2011), we interpreted our niche axes (temperature, moisture, N) as "scenopoetic variables" that have the potential to influence species occurrence but are not themselves directly modified by the presence of the species. This interpretation made good biological sense for *S. purpurea*, which has relatively modest total biomass in its bog habitat and is not a major storage component for N or C (in contrast to *Sphagnum*; Moore et al. 2002). With respect to temperature, moisture, and N, the climatic niche of *S. purpurea* is fairly broad: a large geographical area encompassing relatively dry cool sites in the upper Midwest through relatively warm moist sites in the Florida panhandle.

We input the geographical coordinates, average temperature, annual precipitation, and N-deposition levels for each of the 727 *S. purpurea* locations (§7.1) into MaxEnt. It is customary in the presentation of MaxEnt results to krige the resulting discrete habitat suitability indices onto a continuous heat map of a region or continent, depicting hot spots of high habitat suitability and cool spots of low habitat suitability. However, that depiction was not appropriate for *S. purpurea* because it grows only in specialized, patchy habitats (bogs, poor fens, and seepage swamps). For this reason, we plotted the MaxEnt predictions only for the original 727 occurrence points. This map revealed regions of relatively suitable sites in Minnesota and throughout northern New England, relatively unsuitable sites in the coastal plains of North and South Carolina, and sites of intermediate suitability in the Florida panhandle (figure 7.6).

7.3.2 Comparison of Forecasts of Demographic and MaxEnt Models

The MaxEnt model generated a limited range of habitat suitability indices, from a minimum of 0.12 to a maximum of 0.77, whereas the demographic model generated a full range of persistence probabilities between 0.0 and 1.0, with three distinct modes at 0, 0.55, and 1.0 (figure 7.7). The cluster of MaxEnt probabilities around 0.5 is a documented problem with this SDM framework that is ameliorated with a likelihood-based approach (MaxLike: Royle et al. 2012; Fitzpatrick et al. 2013).

The MaxEnt and demographic projections were not correlated with each other and gave nearly opposite predictions. The MaxEnt habitat suitability indices showed almost no relationship to N levels, either for a model with temperature,

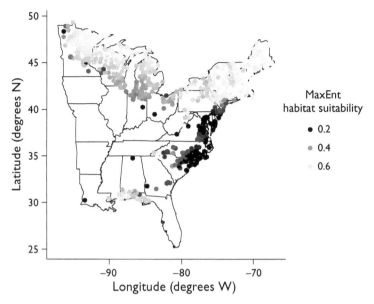

FIGURE 7.6. MaxEnt predictions of habitat suitability (on a scale from 0.0 to 1.0) for *S. purpurea* at 727 locations. The predictions are derived from a default MaxEnt analysis using average annual temperature and annual precipitation from WorldClim, and estimates of N deposition in 2016 from NADP. Light gray points have relatively high habitat suitability, and dark gray points have relatively low habitat suitability. Compare these forecasts with the demographic predictions for the same sites (figure 7.5).

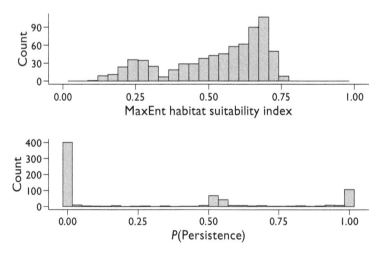

FIGURE 7.7. Histogram of model predictions (*n* = 727 sites). Upper panel: histogram of MaxEnt habitat suitability indices (from figure 7.6). Lower panel: histogram of demographic probabilities of persistence (from figure 7.5).

precipitation, and N (figure 7.8) or for a simpler model that used only N as a predictor variable (and thus was more comparable to the demographic models; figure 7.8). In contrast, the probability of persistence in the demographic models showed a complex, non-linear relationship to N deposition, with a narrow range of intermediate N-deposition rates that were associated with a high probability of persistence and a steep decline in persistence probabilities above this range (figure 7.8). The MaxEnt model predicted good habitat suitability for many sites in the upper Midwest (figure 7.6), whereas the demographic models predicted that many of these sites would be highly vulnerable to extinction (figure 7.5).

7.4 ADDITIONAL FORECASTING SCENARIOS, PAST AND FUTURE

Manipulating N in the field, gathering demographic data to create population projections, and then combining these data with global forecasts for increasing N deposition is much harder than simply compiling species occurrence records, downloading a set of WorldClim variables, and passing the data to MaxEnt (Phillips et al. 2006; Fitzpatrick et al. 2013) or MaxLike (Royle et al. 2012). Our short-term field experiments and long-term monitoring of *S. purpurea* illustrated well the challenges of this kind of modeling and provide new insights into responses of species to ongoing global change.

Our demographic modeling used deposition levels for 2016, the most recent data available at the time we started writing this book. What would the projections have looked like if we used data from other years? The scenarios for the earliest data (1985) and those available from 2000 gave more pessimistic projections, with 75% and 78% of populations, respectively, declining at a relatively rapid rate of >1% per year (table 7.1).

It is interesting to compare these earlier deposition patterns to the history of air-pollution control in the United States. The 1970 Clean Air Act (see chapter 2, §2.1.2) required states to develop plans for meeting air-quality standards by 1977. Several amendments and provisions have been added since its enactment, and the 1990 amendments specifically addressed a reduction in acid rain and the problem of interstate pollution. The latter reflected data showing that NO_x and NH_4 can be transported long distances by weather systems and their associated precipitation. For example, much of the N deposited in New England originates from fossil-fuel consumption for energy production in the upper Midwest and the Ohio River Valley. It is possible that the improved projections for *S. purpurea* demography based on 2016 N-deposition data relative to the models using 1985 and 2000 data reflect the tighter regulations and reductions in air pollution that have occurred since 1990.

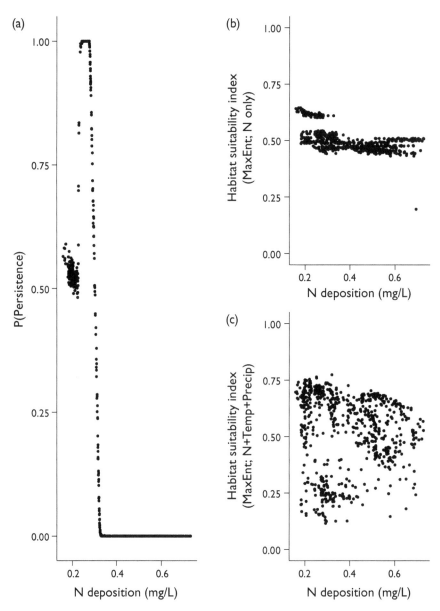

FIGURE 7.8. Demographic and MaxEnt model predictions as a function of 2016 N-deposition rates. Each point represents a *S. purpurea* population site (from figure 7.1). (**a**) Probability of population persistence (from figure 7.5). (**b**) MaxEnt habitat suitability from a model fitted with only 2016 N deposition as the predictor variable. (**c**) MaxEnt habitat suitability from a model fitted with 2016 N deposition, average annual temperature, and annual precipitation (from figure 7.6).

TABLE 7.1. Percentage of *S. purpurea* sites with projected increases ($\lambda \geq 1.00$), slow declines ($0.99 < \lambda < 1.00$), and fast declines ($\lambda \leq 0.99$) as a function of N-deposition levels.

Forecast	1985	2000	2016	+1%	+5%	+10%
Increase	13	6	38	38	34	29
Slow decline	12	16	17	16	18	22
Fast decline	75	78	45	46	47	49

Three hindcast scenarios are shown for the years 1985 (the first year of NADP N-deposition data), 2000, and 2016. Three forecast scenarios are shown for constant increases of 1%, 5%, and 10% above N deposition values in 2016.

However, atmospheric N deposition in the United States may start increasing again in the future due to drought and increasing fire frequency (Westerling et al. 2003), increased consumption of fossil fuels (Davis and Caldeira 2010), and the rolling back of federal air-quality and fuel-economy standards (Hogrefe et al. 2004).[4] Table 7.1 shows the forecasted effects on *S. purpurea* population growth rate (λ) of increasing N deposition levels by constant amounts of 1%, 5%, and 10% above the baseline level of 2016. Unsurprisingly, these increases lead to modest increases in the proportions of *S. purpurea* populations that are declining rapidly.

7.5 SYNTHESIS

Although species distribution modeling based on presence-only data continues to grow in popularity and statistical complexity, there have been few empirical comparisons of the performance of MaxEnt with models that are based on other kinds of data. For models of *S. purpurea*, the comparisons were not at all similar, and reflected three broad classes of problems with the current species distribution modeling paradigm.

The first problem is the statistical performance of the MaxEnt algorithm does not generate comparable results with a likelihood-based model. The second problem is that mid-latitude terrestrial plants and animals already experience a vast amount of seasonal change in temperature and precipitation, so they may not be as vulnerable to climatic change as high-latitude species that are tied to specialized habitats or low-latitude species that may already be close to their physiological maximum thermal tolerance. Although *S. purpurea* is a habitat specialist, it occupies a very broad climatic niche, can be easily introduced and established outside of its range, and probably harbors substantial genetic variation within and among subpopulations that may allow it to adapt successfully to climatic change. For organisms like

S. purpurea, changes in other environmental drivers, such as nutrient deposition and land use, may be the critical determinants of population extinction or persistence.

The third problem is that even when we include N deposition in an SDM, MaxEnt fails to detect its importance. This is because the MaxEnt predictions are based solely on the environmental conditions where the species occurs and where pseudo-absences are imputed. MaxEnt cannot account for either demographic performance or genetic variation of populations at these sites. We are sympathetic to the goals of species distribution modeling, the urgency of predicting how species might respond to a changing climate, and the potential utility of compiling species occurrence records to assist in this analysis. But ecologists need to get back to basics. We need to get into the field; do reciprocal transplant and common garden experiments; and measure rates of survivorship, growth, and reproduction in response to changing environmental drivers. We also need to measure genetic variation in these populations and understand the potential for local adaptation and phenotypic plasticity as responses to climatic change. These data are not as easy to gather as occurrence data downloaded from GBIF and WorldClim. But they can certainly be gathered in the framework of a dissertation or 3- to 5-year grant.

With empirical data in hand, GBIF records can be put to better use. Empirical data will allow us to expand more confidently the spatial and temporal scales of our forecasting efforts to places where species already occur. There may still be great uncertainty in experimental results, depending on which variables were manipulated and what treatment levels were used. But at least our forecasts will not be misled by a simple mapping of species occurrences on only two measured climatic variables.

Part III

Ecology of the *Sarracenia* Community

CHAPTER EIGHT

Context

Community Ecology, Community Ecologies, and Communities of Ecologists

> *This frugal Community* [of insects] *are wisely employed in providing for Futurity, and collecting a copious Stock of the most balmy Treasures.*
> —James Hervey (1747: 108)

Although the parts of this book are ordered from small (physiology) to large (ecosystems) scales of biological organization, we began our research with pitcher plants in the middle, focusing on the ecology of the community of bacteria, protists, and arthropods—collectively called "inquilines"—that live within their water-filled pitchers (appendix A, §A.3.2). In our first grant to study *Sarracenia purpurea* and its inquiline food web, we wrote that the data we collected would *enable the development, refinement, and testing of a mathematical model of community assembly in a dynamic habitat.* Our goal was to adapt null models (Gotelli and Graves 1996) and Markovian succession models (Horn 1975) for equilibrium communities at a single trophic level to the assembly of multitrophic food webs (Drake et al. 1993). This goal was challenging because species co-occurrence, succession, and food web or network analysis represent three disparate research lineages in commumity ecology.[1]

We thought that we could accomplish our goal in a normal 5-year grant cycle, but it has taken us more than 20 years just to accumulate the necessary data and integrate existing and new theory to start working on community assembly in a dynamic habitat. The rest of this chapter gives a brief overview of how ecologists have defined communities, the utility of space–time substitutions, and the importance of networks.

Chapter 9 begins with a description of the food web of *S. purpurea* inquilines (with more elaboration in appendix A, §A.3.2; figures 1.2 and A.7; see also Miller et al. 2018). We then analyze quantitative patterns of species co-occurrence, temporal succession, and food-web organization of *Sarracenia* inquilines. Apparently, the inquilines read neither Gotelli and Graves (1996) nor Horn (1975), and analyses of species co-occurrence and succession were not as useful in explaining pattern as they are in other assemblages. In chapter 10, we illustrate how the *Sarracenia* food web scales with respect to other published food webs and discuss its utility as a model experimental system. This discussion leads to suggestions of how we might weave together, at least conceptually, the various ways of considering ecological communities in space and time.

8.1 BACKGROUND

8.1.1 What Is an Ecological Community?

Community ecologists describe, analyze, model, and experimentally manipulate ecological communities, but ecologists are vague about what they mean by communities.[2]

Communities may be defined by direct pairwise interactions between species [consumers sharing a limited resource, herbivores and their plants, pollinators and their "balmy nectar" (figure 8.1), predators and their prey, and parasites and their hosts; Weiher and Keddy 1999]. Communities may occur as "replicated patches" in similar environments (Diamond 1986); vary along environmental gradients (Maestre et al. 2009); or change with other explicitly spatial variables such as latitude, elevation, or depth (ter Braak and Prentice 1988). Location (place, space) is considered to be an essential part of an ecological community, but time is not, perhaps reflecting the deeply held view that the natural world is in a balanced equilibrium (cf. Kricher 2009; Ellison 2013). This temporal myopia of (animal) community ecologists has been noted before (e.g., Huston 1984) but is being overcome with the recent development of methods for analysis of time-varying "multilayer" networks (e.g., Li and Li 2016; Pilosof et al. 2017; López et al. 2018; Hutchinson et al. 2019). In contrast, plant ecologists studying succession have long appreciated the temporal dimension of community structure, but whether the assemblages they study can be properly called communities has been debated for more than a century (Gleason 1926; Tansley 1935; Clements 2019; Liautaud et al. 2019). Operationally, MacArthur's (1971) definition still may be the most useful: "[a]ny set of organisms currently living near each other and about which it is interesting to talk."

8.1.2 Substituting Space for Time, and Vice Versa

Studies of species co-occurrence, food webs and other networks, and succession continue to dominate contemporary community ecology (Stroud et al. 2015), perhaps because they each emphasize the importance of biotic interactions in

FIGURE 8.1. The first mention of an ecological community is in James Hervey's (1746) book *Reflections on a Flower-Garden*, bound together with his *Meditations Among the Tombs*. Hervey is shown here, with his back to us, in this pen-and-watercolor drawing by William Blake. (William Blake, *Epitome of James Hervey's Meditations among the Tombs*, c. 1820–1825. N 02231 / B 770; Pen and watercolour 431 × 292 mm ($16\frac{15}{16} \times 11\frac{1}{2}''$). Image ©Tate.

producing pattern in natural communities. However, it is much easier to collect spatial replicates for co-occurrence analysis or to construct species-interaction networks than it is to collect temporal replicates to document succession. Indeed, because it is so difficult to directly track communities through long periods of time (Dornelas et al. 2018), space-for-time substitutions dominated studies of ecological change until the advent of the Long-Term Ecological Research (LTER) program in 1980 (Willig and Walker 2016).

For example, classic succession models of facilitation were constructed from chronosequences of shoreline vegetation (e.g., Cowles 1899; Clements 1916; Fastie 1995). However, direct tests of the assumption that space can be substituted for time have not been well supported for several classic chronosequence studies (Johnson and Miyanishi 2008; but see Blois et al. 2013 for a Holocene-scale perspective). In a parallel use of space-for-time substitutions, species distribution modelers regularly use spatial patterns of species occurrences to forecast (e.g., chapter 7, §7.3) and hindcast changes in species and community composition associated with environmental change (Record et al. 2013).

Conversely, the few studies that have examined how species co-occurrence patterns change through time (a time-for-space substitution) also have yielded inconsistent results at vastly different temporal scales. At very small temporal scales, *S. purpurea* inquilines showed a weak increase in aggregation within a single season (see chapter 9, §9.3, and Ellison et al. 2003). At the other temporal extreme, the species identity of aggregated and segregated species pairs has been inconsistent when considered across the entire Quaternary Period (Blois et al. 2013) or even since the Carboniferous (Lyons et al. 2016).

Of course, the interaction of space and time has not been ignored by community ecologists. For example, mechanistic models of forest succession incorporate the distance and identity of neighboring individuals (e.g. Pacala et al. 1996), and metacommunity models (Leibold and Chase 2017) have taken into account patch structure and dispersal potential. There also are decades of studies on spatial variation (e.g., Zajac and Whitlach 1982; Farrell 1991), spatial scale (e.g., Zajac et al. 1998), environmental gradients (e.g., Letcher et al. 2015), and disturbance regimes (e.g., Platt and Connell 2003) on the timing and outcome of successional sequences in many habitats.

Still, most studies of succession and species co-occurrence continue to focus on taxonomic guilds (sensu Root 1967; see also Stroud et al. 2015): sets of related species with broadly similar life histories that potentially interact directly and indirectly. There are good theoretical and practical reasons for restricting the analyses to taxonomic guilds. These species often share a common evolutionary history, similar life histories, or resource requirements, so they are expected to interact, especially through the use of shared resources of food, space, or time (Schoener 1974). From a practical standpoint, the dual challenges of taxonomic identification and standardized, quantitative sampling methods also have led many ecologists to restrict their studies and samples to sets of related species.

8.1.3 The Importance of Networks

The continued focus on studies of closely related species has somewhat isolated studies of succession and co-occurrence from the other major thread of community organization: food webs and networks. Many of the strongest ecological interactions—parasitism, predation, pollination, and mutualism—occur across trophic levels and between distantly related taxa. Evidence for these trophic interactions often is stronger than the evidence for competition within trophic levels, which can be inferred only indirectly from patterns of species co-occurrence or segregation (Strong et al. 1984; Gotelli and McCabe 2002).

Long before it became fashionable for the study of human social and computer networks, the importance of networks was appreciated by food-web ecologists (Elton 1927). But early studies of food webs were crippled by a lack of data, the construction of composite webs that lumped functionally equivalent or difficult-to-identify taxa (e.g. "algae"), ignorance of spatial or temporal variation in food-web structure, and a failure to include measurements of interaction strength (Paine 1988; McCann 2012). More recently, there has been a renewed appreciation of these sources of variation and that network structure is changes in time and varies in space. The identification of dynamic, "multilayer" networks and the development of methods and tools to analyze them (e.g., Li and Li 2016; Pilosof et al. 2017; Hutchinson et al. 2019) are rapidly advancing our understanding of food-web structures and dynamics. These methods and tools have allowed us to come much closer to achieving our goal of understanding community assembly in dynamic habitats.

8.2 NEXT STEPS

Community ecologists seek to identify processes driving spatial and temporal patterns of species associations. Our work has woven together ideas and insights from three major research themes in community ecology: species co-occurrence, food webs and other networks, and succession. In the next chapter, we bring together these themes in a model of the temporal dynamics of the structure and interactions of the inquiline food web. In chapter 10, we place these data and results in the context of general syntheses of food-web structure and discuss the utility of the inquiline food web as a model experimental system for exploring basic and applied questions in the structure, function, and management of food webs.

CHAPTER NINE

The Small World

Structure and Dynamics of Inquiline Food Webs in Sarracenia purpurea

> *One finds in a single body of water... a little world within itself—a microcosm within which all the elemental forces are at work and the play of life goes on in full, but on so small a scale as to bring it easily within the mental grasp.*
> —Stephen A. Forbes (1877: 77)

9.1 COMPOSITION AND STRUCTURE OF THE *SARRACENIA PURPUREA* FOOD WEB

9.1.1 The Inquilines

The *S. purpurea* food web is an aquatic, detritus-based ("brown") food web. The input to this donor-controlled food web is (primarily) insect prey captured by the plant that drowns in the liquid-filled pitchers (Butler et al. 2008; Ellison and Gotelli 2009; Miller et al. 2018). The living components of the food web—the inquilines—include one species that completes its entire life cycle within the pitcher (the mite *Sarraceniopus gibsoni*); three species that live in pitchers only during their larval stages (the midge *Metriocnemus knabi*, the mosquito *Wyeomyia smithii*, and the sarcophagid fly *Fletcherimyia fletcheri*); and rotifers, protozoa, and microbial taxa that live in both pitcher fluid and a variety of other aquatic habitats (appendix A, §A.3.2).

In most of the analyses we discuss in this chapter, we use only the five invertebrate species in the "macrobial food web": (1) the voracious and cannibalistic top-predator larvae of *F. fletcheri*; (2) the filter-feeding larvae of *W. smithii*; (3) the detritivorous larvae of *M. knabi*; (4) the common and abundant filter-feeding bdelloid rotifer *Habrotrocha rosa*; and (5) the particulate organic matter (POM) feeding mite *S. gibsoni*. These taxa are the ones that can be seen by eye or with a 10× hand-lens in the field.

In our own field experiments, we rarely collected data on the hundreds of operational taxonomic units (OTUs) of protozoa, bacteria, viruses, yeast, fungi, and algae that also are members of this aquatic food web. However, in observational studies in the field and in laboratory experiments, we and our colleagues have collected data at small and large geographical scales on the more complete (macrobial + microbial) food web of *S. purpurea* (Buckley et al. 2003, 2004, 2010; Peterson et al. 2008; Baiser et al. 2012; Paise et al. 2014; Canter et al. 2018; Boynton et al. 2019; Bittleston et al. 2020).

9.1.2 Network Structure of the Sarracenia purpurea Food Web

The *S. purpurea* food web has been used extensively as a model in studies of keystone predation and other top-down effects (Addicott 1974; Cochran-Stafira and von Ende 1998; Kneitel and Miller 2002; Gotelli and Ellison 2006a; Hoekman 2007, 2010, 2011; Peterson et al. 2008; Hoekman et al. 2009); omnivory and other intermediate-level trophic interactions (Kneitel 2007); bottom-up effects (Bradshaw and Creelman 1984; Heard 1994; Gotelli and Ellison 2006a; Hoekman 2007, 2010, 2011; Hoekman et al. 2009); small- and large-scale biogeography (Harvey and Miller 1996; Buckley et al. 2003, 2004, 2010; Baiser et al. 2012; Gray et al. 2016); and metacommunity organization (Miller and Kneitel 2005; Baiser et al. 2013). However, the definition and characterization of the "*S. purpurea* food web" have varied considerably among these studies (figure 9.1; table 9.1).

The number of identified taxa included in the *S. purpurea* food web has ranged from as few as 4 (Hoekman 2010) to as many as 64 (Baiser et al. 2012, whose full web also included an additional 26 morphotaxa assigned only to phyla). All canonical webs (figure 9.1) have included captured prey, the common illoricate rotifer *Habrotrocha rosa*, and the four obligate macrobial taxa (table 9.1). Inclusion or exclusion of microbes, and the degree to which microbial taxa have been aggregated into taxonomic groups including bacteria, yeasts, fungi, or protozoa, have depended on the question or hypothesis of interest and increasing ability to identify microbial OTUs (Peterson et al. 2008; Paise et al. 2014; Canter et al. 2018; Boynton et al. 2019). For example, experimental studies aimed at testing whether *W. smithii* is a keystone predator frequently have used simplified food webs consisting only of multiple species of competing protozoa and the mosquito that feeds on them (Addicott 1974; Cochran-Stafira and von Ende 1998; Kneitel and Miller 2002; Hoekman 2007, 2010). Similarly, models (Mouquet et al. 2008) and experiments (Butler et al. 2008) quantifying nutrient or energy fluxes in the *Sarracenia* food web have segregated the macrobes but pooled the microbes into two (bacteria, protozoa + rotifers) or three (bacteria, protozoa, rotifers) taxonomic groups.

At the other extreme, for our continental-scale assessment of *S. purpurea* inquilines, we identified several additional macrobes, including nematodes and

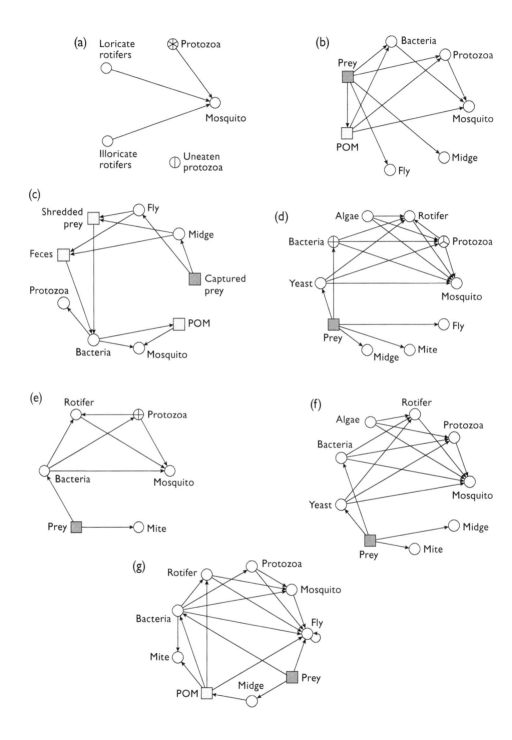

tardigrades, and dozens of genera and species of protozoa and culturable bacteria (appendix A, figure A.8; Buckley et al. 2003, 2010; Baiser et al. 2012). Not surprisingly, by disaggregating protozoa and bacteria into many individual taxa, the effective linkage density and ascendency[1] were an order of magnitude greater ($LD_E = 21.15$, $\log_{10}(A) = 2.81$, respectively) than that of our more aggregated nine-taxa web ($LD_E = 2.69$, $\log_{10}(A) = 1.37$, respectively; table 9.2). Connectedness and trophic depth, however, were similar between the aggregated and disaggregated webs ($C = 0.22$ and $TD_E = 2.49$ for the web shown in appendix A, figure A.7 vs. $C = 0.19$ and $TD_E = 1.77$ for the disaggregated web of Baiser et al. 2012).

9.2 CO-OCCURRENCE ANALYSIS OF *SARRACENIA PURPUREA* INQUILINES

A simple investigation of the structure of the *Sarracenia* community would begin with a co-occurrence analysis (appendix E, §E.1). Note that co-occurrence analyses normally are done on "communities" of species that interact within a trophic level or guild (e.g., trees, seed-eating birds [Root 1967; Stroud et al. 2015]), but there is no reason that the same kinds of analyses cannot be applied to a community defined more broadly to include different functional groups (e.g., all New Hebridean birds, as in Diamond 1975) or trophic levels, such as the *S. purpurea* food web (Ellison et al. 2003).

9.2.1 Quantifying and Testing Inquiline Co-occurrence

We tested whether the patterns of species co-occurrence in the *Sarracenia* food web were statistically random, aggregated, or segregated. We used data from an

---

FIGURE 9.1. Seven canonical *Sarracenia* food webs from: (**a**) Addicott (1974); (**b**) Bradshaw and Creelman (1984); (**c**) Heard (1994); (**d**) Cochran-Stafira and von Ende (1998); (**e**) Kneitel and Miller (2002) with protozoa from Kneitel (2007); (**f**) Mouquet et al. (2008); and (**g**) Butler et al. (2008) and modified from figure A.7. Each web is represented as a network diagram, with arrows going *from* prey, detritus, etc. *to* predators, transformed detritus, etc. Captured prey is represented by a dark gray square, and its immediate or transformed products (variously shredded prey, particulate organic matter [POM], or feces) are represented by light gray squares. Consumers are represented by circles. For individual taxa within a group of consumers that were identified separately by the investigators, the number of identified taxa are represented by the number of "slices" in the circular pie (e.g., three genera of protozoa identified by Cochran-Stafira and von Ende 1998). See table 9.1 for the scientific names for each of the taxa and table 9.2 for metrics of network and food-web structure for each of these webs.

TABLE 9.1. Common and unique taxa identified in canonical food webs of
Sarracenia purpurea inquilines illustrated in figure 9.1.

Author[a]	Common name	Scientific name(s)[b]
All webs	Rotifer	*Habrotrocha rosa*
	Mite	*Sarraceniopus gibsoni*
	Midge	*Metriocnemus knabi*
	Mosquito	*Wyeomyia smithii*
	Fly	*Fletcherimyia fletcheri*[c]
Addicott (**a**)	Protozoa	*Bodo* spp.
		Colpidium campylum
		Colpoda steinii
		Cyclidium elongatum
		Cyrtolophosis elongata
		Urotricha ovata
	Uneaten protozoa	*Monas* spp.
		Vorticella sp.
Cochran-Stafira and von Ende (**d**)	Protozoa	*Bodo*
		Colpoda
		Cyclidium
Kneitel and Miller (**e**)	Protozoa	*Bodo*
		Colpoda
		Colpidium
		Poterioochromonas

[a]Canonical *S. purpurea* food webs from Addicott (1974); Bradshaw and Creelman (1984); Heard (1994); Cochran-Stafira and von Ende (1998); Kneitel and Miller (2002); Mouquet et al. (2008), and our own work (figure A.7; see also Butler et al. 2008). Boldfaced letters correspond to network diagrams with subdivided consumers (i.e., pies with slices) in figure 9.1.

[b]Nomenclature checked in the 2019 Catalogue of Life (http://www.catalogueoflife.org/) and the World Ciliate Catalog (https://www.zobodat.at/).

[c]*Blaesoxipha fletcheri* in earlier papers.

experiment in which we repeatedly monitored the assembly of the *S. purpurea* macrobial food web. Weekly throughout the 1999 growing season at both Hawley and Molly Bogs, we nondestructively censused the inquilines in the first new pitcher of 40 plants chosen randomly at each site (Ellison et al. 2003). Because co-occurrence analysis assumes that the community has reached equilibrium, we show the results only for pitchers sampled on July 20, 1999 (the complete time series was reported by Ellison et al. 2003). This sampling date is appropriate because by midsummer, all macroinvertebrates that are likely to colonize the pitchers have already done so and last-instar *Fletcherimyia* have not yet left the pitchers to pupate in the surrounding *Sphagnum*. However, even at this date, not all pitchers were colonized by macroinvertebartes. On July 20, 1999, only 20

TABLE 9.2. Metrics of network and food-web structure of canonical food webs of *Sarracenia purpurea* inquilines illustrated in figure 9.1.

Author[a]	N	L	C	LD	LD_E	TD_E	A
Addicott (**a**)	11	8	0.07	0.73	2.83	1.00	0.00
Bradshaw and Creelman (**b**)	7	10	0.20	1.43	2.39	1.75	8.10
Heard (**c**)	9	12	0.15	1.33	1.72	4.06	24.26
Cochran-Stafira and von Ende (**d**)	10	17	0.17	1.70	2.90	2.02	17.25
Kneitel and Miller (**e**)	6	8	0.22	1.33	1.96	2.09	8.49
Mouquet et al. (**f**)	9	15	0.19	1.67	2.72	2.02	15.22
Ellison and Gotelli (**g**)	9	18	0.22	2.00	2.69	2.49	23.67

See table E.4 in appendix E for definitions and calculations of metrics.

[a]Canonical *S. purpurea* food webs from Addicott (1974); Bradshaw and Creelman (1984); Heard (1994); Cochran-Stafira and von Ende (1998); Kneitel and Miller (2002); Mouquet et al. (2008), and our own work (appendix A, figure A.7; see also Butler et al. 2008). Boldfaced letters correspond to network diagrams in figure 9.1.

(of 40) pitchers at Hawley Bog and 21 (of 40) pitchers at Molly Bog contained at least one macroinvertebrate taxon (table 9.3). The remaining "empty" leaves either had not accumulated any water and hence were unsuitable for oviposition, colonization, or establishment, or were too small at that point in the season to support a macrobial assemblage.

Methods We used the C-score (appendix E, equation E.1; Stone and Roberts 1990) to quantify co-occurrence patterns because it has good statistical power for detecting patterns of nonrandomness (Gotelli 2000). Larger values of this index indicate that species pairs, on average, are more segregated and co-occur in relatively few sites. We tested whether observed patterns were significantly different from random using three null models (Gotelli 2000), each of which reflects different assumptions about the distribution of inquilines among pitchers (appendix E, §E.1.3).

First, a **fixed–equiprobable** null model preserved the overall relative abundance of the inquilines in the simulations but allowed the number of species per pitcher to vary freely and randomly. This choice seemed reasonable for comparing inquiline co-occurrence among a set of identically aged leaves. Second, a **fixed–proportional** null model set the probabilities of occurrence of species in a pitcher based on some measure of site suitability. In this case, we used the measured volume of each leaf as a proxy for site size. All other things being equal, we expected occupancy by all species to increase with increasing water volume (Gotelli and Ellison 2006a). Finally, a **fixed–fixed** null model was used for completeness, but the small number of species used in this analysis would be expected to lead to an unacceptably large Type II error rate (Ulrich and Gotelli 2007a).

We compared the C-scores of the observed Hawley and Molly Bog co-occurrence matrices (table 9.3) to the distribution of C-scores generated by 10,000 simulations of each of the three null models. To ensure repeatability of the results, we reran the simulations ($n = 10,000$) and analyses starting with a different random-number seed. Estimated P values were consistent to two decimal places. An α-level equal to 0.05 was used as the upper-bound of a type I error, the probability of incorrectly rejecting the true null hypothesis (Kuffner and Walker 2019).

Results   Figure 9.2 illustrates the results for the Molly Bog matrix (table 9.3) tested with the fixed–fixed algorithm. The observed C-score for the Molly Bog matrix was 22.4, and the average C-score from 10,000 simulated null matrices was 21.5. Of these 10,000 simulated C-score values, 9,195 of them were less than the observed C-score of 22.4, 172 = 22.4, and 633 > 2.4. Thus, the two-tailed probability of observing a C-score $\geq$ 22.4 with the null distribution generated from the fixed–fixed algorithm was $(172 + 633)/10,000 = 0.08$. Although the average pairwise segregation of species co-occurrences was larger than expected, it was not "significant" given $\alpha = 0.05$. For the Hawley Bog matrix tested with the fixed–fixed algorithm, the observed C-score was 6.6; $P = 0.66$.

The other two algorithms also suggested that the Molly Bog co-occurrence data were not different from random expectation, but the fixed–proportional model revealed significant segregation for the Hawley Bog matrix (figure 9.3). Which of these results was "correct"? The answer should be based on the validity of the biological assumptions underlying each model, rather than the one that gave a result of $P < 0.05$.

From our perspective, the fixed–equiprobable model was the most appropriate for these data. With only five taxa, it seemed unrealistic to constrain the column totals as in the fixed–fixed algorithm, and with such a small number of taxa, the type II error probability was likely to be inflated. We also saw no reason why each of the experimental pitchers—assuming they already had at least one taxon present—could not have supported all five taxa in a real or modeled assemblage. Although the leaves may have differed in their suitability, we are skeptical that the differences were as large as might be predicted by water volume (see chapter 10) and would be implied by the fixed-proportional null model.

For example, in the Hawley data, leaf #61 had a water volume of 9.50 mL, and leaf #65 had a volume of 34.0 mL. In fixed–proportional null-model simulations, this would mean that the probability of occurrence for all species was $34/9.50 = 3.6$ times greater for leaf #65 than for leaf #61. But both of these leaves contained the same three taxa. Pitcher water volume is strongly associated with both inquiline

TABLE 9.3. Co-occurrence matrices for Molly Bog and Hawley Bog. Each row represents a *S. purpurea* inquiline taxa.

Molly Bog

											Plant number										
	11	12	15	16	19	110	116	118	119	120	41	43	44	49	410	411	412	413	414	418	419
Fletcherimyia	0	0	0	0	0	0	0	0	1	0	0	0	0	0	1	0	0	0	0	0	0
Habrotrocha	1	0	0	1	0	1	1	1	0	1	1	1	1	0	0	1	0	1	1	1	1
Metriocnemus	0	0	0	0	0	1	0	1	0	0	0	0	0	1	0	1	1	0	0	1	0
Wyeomyia	0	0	0	1	0	1	0	0	0	0	1	1	1	1	0	0	0	0	1	1	0
Sarraceniopus	0	0	0	0	1	0	1	1	0	1	1	1	1	0	0	1	0	0	1	1	1

Hawley Bog

										Plant number										
	1	4	8	9	12	13	17	32	36	48	52	57	61	64	65	68	72	73	76	80
Fletcherimyia	0	0	0	0	0	0	1	0	0	0	0	0	0	0	0	0	0	0	0	0
Habrotrocha	1	1	1	1	1	1	1	1	1	1	1	1	1	1	1	1	1	1	1	1
Metriocnemus	0	1	0	1	0	0	1	0	1	1	0	0	1	0	1	1	0	0	1	1
Wyeomyia	0	0	1	0	0	0	1	1	0	1	0	1	1	0	1	1	1	1	0	1
Sarraceniopus	0	1	1	1	1	1	1	0	0	1	1	1	0	1	0	1	1	0	1	0

Each column represents the contents of a single leaf containing 1 or more taxa from a cohort of 40 monitored plants. Entries indicate the presence (1) or absence (0) of a taxon in a leaf censused on July 20, 1999.

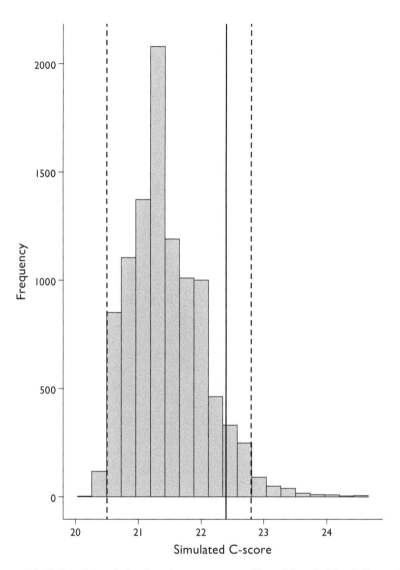

FIGURE 9.2. Null-model analysis of species co-occurrence. The solid vertical line indicates the observed C-score (22.4) for the Molly Bog co-occurrence matrix (table 9.3). The histogram illustrates the distribution of 10,000 C-score values generated from a fixed–fixed randomization of the matrix. The dashed vertical lines encompass 95% of the simulated values.

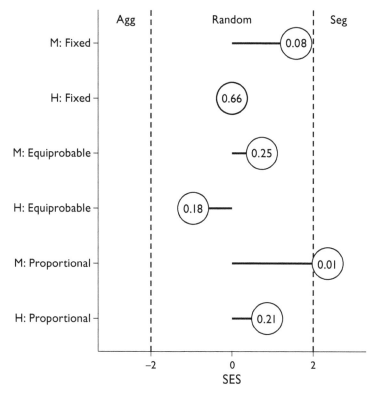

FIGURE 9.3. Null-model analyses of species co-occurrence at Molly Bog (M) and Hawley Bog (H). Fixed–fixed, fixed–equiprobable, and fixed–proportional are different algorithms for randomizing the data. Values on the x-axis are the standardized effect sizes (SES), a metric derived from the null-model analysis (figure 9.2) that summarized the tendency toward aggregation, segregation, or randomness in pairwise species associations. The number in each circle is the one-tailed P value from the null-model analysis. Data are from table 9.3.

(co)occurrence and abundance and with the capture and accumulation of prey (chapter 10; Gotelli and Ellison 2006a). However, we think that colonization probabilities were tied more closely to the effective surface area of water at the top of the pitcher than to the volume of the entire leaf.

This experiment also included treatments in which we delayed the time at which pitchers were opened and available for colonization (Ellison et al. 2003). Across all the treatments, there was little deviation from randomness when tested with the fixed–equiprobable model (Ellison et al. 2003). However, the assemblages at Hawley Bog did become more aggregated in co-occurrence as the season progressed

(Ellison et al. 2003), as might be expected when a food web assembles from scratch (Drake 1990).

Nevertheless, we could argue for the existence of a classic assembly rule of the sort described by Diamond (1975), albeit arising here from predation, not competition. In the *Sarracenia* microecosystem, the top predator *F. fletcheri* consumes *W. smithii* and *M. knabi* larvae and other macroinvertebrates. At Molly Bog, there were three pitchers (#12, #119, #410) that contained *F. fletcheri* but no other invertebrate taxa. However, this same checkerboard pattern was not observed at Hawley Bog, where among our experimental plants there was only a single occurrence of *F. fletcheri*—and that in a pitcher with all of the other four taxa.

In sum, with the possible exception of *F. fletcheri* at Molly Bog, the *S. purpurea* macroinvertebrate assemblage shows little evidence of nonrandom co-occurrence between taxa. Given the limited number of taxa involved, the homogeneity of the newly opened leaf habitat, and the diverse body sizes and trophic roles of the taxa, this result should not be surprising.

9.3 SUCCESSION OF THE INQUILINE FOOD WEB

Because we censused the *S. purpurea* food web not only in space (figures 9.2 and 9.3 and appendix A, figure A.8) but also through time, we could compare its temporal changes to predictions of a Markovian successional model (appendix E, §E.3.1; Horn 1975). Depending on the interactions between the different stages and their modifications of the environment, successional sequences can be modeled as being caused by facilitation, inhibition, or tolerance (Connell and Slatyer 1977). An important prediction of the classic facilitation model is that assemblages that are initially different will converge to the same composition through time. But the snapshot view of individual *S. purpurea* leaves (table 9.3) did not match that prediction: at both sites, replicate leaves of the same age in midsummer had different numbers and composition of species.

Facilitation, inhibition, or tolerance models of succession also require discrete "community types" or stages through which assemblages may pass (appendix E, §E.3). Operationally, it was not possible with an assemblage of only five macrobial taxa to classify multispecific assemblages into different stages. However, if the interactions between the individual taxa are strong, as would be expected in trophic webs with few species, each taxon could be considered as its own "stage" in a successional model (Horn 1975). By using such a classification, we predicted that some species would appear at different times in a temporal sequence. However, this prediction also did not match our observations of the inquiline assemblages within *S. purpurea* pitchers (figure 9.4).

TABLE 9.4. Pairwise temporal correlations of *S. purpurea* inquiline incidence.

	Fletcherimyia	*Habrotrocha*	*Metriocnemus*	*Sarraceniopus*	*Wyeomyia*
Fletcherimyia	∅	**−0.88**	**−0.65**	−0.28	**−0.66**
Habrotrocha	−0.35	∅	0.49	0.42	**0.69**
Metriocnemus	−0.02	**0.69**	∅	−0.15	0.37
Sarraceniopus	**0.67**	−0.16	0.02	∅	0.17
Wyeomyia	**−0.81**	**0.57**	0.44	**−0.67**	∅

Each entry is the Spearman rank correlation coefficient of temporal incidence (figure 9.4) between a given pair of taxa. Boldfaced entries are statistically significant ($P < \alpha = 0.05$). Entries above the diagonal are for Molly Bog, below the diagonal for Hawley Bog.

We plotted the incidence of each taxon as a function of time in the season (figure 9.4). The patterns of incidence were similar at both sites: all taxa except *F. fletcheri* could be found with some frequency throughout the growing season. Peak incidences for *H. rosa*, *S. gibsoni*, and *M. knabi* occurred 15–60 days later at the more northern Molly Bog site. These temporal shifts were consistent with a universal scaling model in which ectotherm development depends on temperature and reflects the underlying biochemical kinetics of metabolism (the "10°C rule": Charnov and Gillooly 2003).

Although there was no evidence of successional sorting from either the spatial (figure 9.3) or temporal (figure 9.4) data on species replacements, we did note that the abundance of the top predator *F. fletcheri* peaked early in the season and usually completed its life cycle by midsummer. As its populations declined through time, the abundances of the other taxa mostly increased. Simple pairwise correlations of incidence did identify many statistically significant negative associations between *F. fletcheri* and the other four taxa, with the patterns being a bit stronger at Molly Bog (table 9.4).

Like other top predators, *F. fletcheri* might have caused the local extinction of other taxa within single leaves. However, an alternative hypothesis is that the association was coincidental and simply reflected the low abundances of most taxa at the start of the season, when dispersal may have been limited, and when tiny eggs or first instars of *W. smithii* or *M. knabi* were not easily detectable in field censuses. The positive correlations between *W. smithii* and both *M. knabi* and *H. rosa* also were consistent with Heard's (1994) identification of a processing-chain commensalism in the *S. purpurea* food web. However, only experimental manipulations of inquilines from newly opened pitchers could reliably distinguish between hypotheses of trophic interactions and phenological differences between species (Heard 1994; Kneitel and Miller 2002; Kneitel 2007; Gray 2012; Miller and terHorst 2012; Canter et al. 2018).

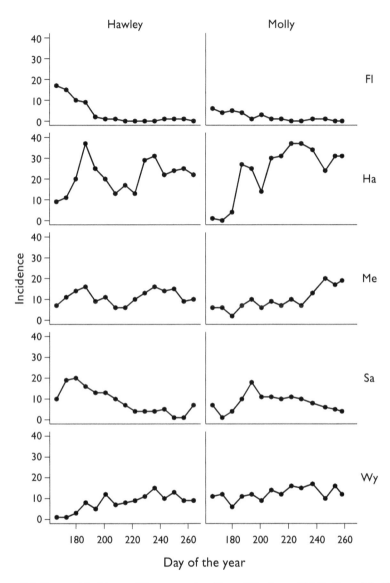

FIGURE 9.4. Temporal incidence of *S. purpurea* inquilines at Hawley and Molly Bogs, July 1999. Each panel represents a different taxon: Fl = *Fletcherimyia fletcheri*; Ha = *Habrotrocha rosa*; Me = *Metriocnemus knabi*; Sa = *Sarraceniopus gibsoni*; and Wy = *Wyeomyia smithii*. The *y*-axis (Incidence) identifies the number of occupied leaves out of 40 that were repeatedly censused. The data in figure 9.3 correspond to day-of-year 201 in this figure.

9.4 DYNAMICS OF THE *SARRACENIA PURPUREA* FOOD WEB

The temporal patterns of species associations that we observed in the *Sarracenia* macrobial food web could not be explained easily by co-occurrence analysis, static metrics of inquiline food-web structure, or classic models of species succession. Although functioning food webs developed in any water-filled pitcher, spatially and temporally replicated assemblages exhibited co-occurrence patterns that were largely random (figure 9.3; Ellison et al. 2003). Pitchers are empty (and likely sterile) before they open (Hepburn and St. John 1927) but the inquilines (and microbes) colonize rapidly once they open. Yet the temporal sequence of species occurrence (figure 9.4; Ellison et al. 2003) did not conform to any of the classic succession pathways. In short, tools designed to analyze co-occurrence and inter-actions *within* a single trophic levels did not capture much of the complexity of the structure of a multitrophic food web (Olesen et al. 2011; Cirtwill et al. 2019). In this section, we examine the structure and function of the inquiline community of *S. purpurea* as a dynamic food web: a "multilayer network" (sensu Hutchinson et al. 2019) in which the additional layer is time.

9.4.1 Temporal Changes in Food-Web Structure

We used the data from the 1999 sampling at Hawley and Molly Bogs described earlier (table 9.3), but in this case we used data on both occurrence (presence or absence) and abundance of each taxon (figure 9.4).

First, we used the occurrence data for each pitcher at each sampling date to compute the different metrics of network and food-web structure described in table (9.2). We used the food web illustrated in figure 9.1g as the most complete network possible. For each replicate pitcher at each sample date, we assigned a value of 1 to each node if the species was present and a 0 if it was absent.

At both bogs, all measures of network and food-web structure were lowest at the beginning of the season as the pitchers were being colonized (figure 9.5). At Hawley Bog, number of links, connectance, linkage density, and trophic depth then increased to a peak in July but declined after *F. fletcheri* pupated in midsummer. There was some indication thereafter of a recovery of these measures as *H. rosa* and *W. smithii* increased in abundance (figure 9.4). *Fletcherimyia fletcheri* consumes available *H. rosa* rapidly (Błędzki and Ellison 1998) and loss of this predator increased the likelihood we would detect *H. rosa* in our samples. In a warm year, Hawley Bog is at the latitude at which *W. smithii* may produce two generations per year (Bradshaw and Holzapfel 2001); the second generation escapes some of the predation pressure from *F. fletcheri* experienced by the first generation.

In contrast, ascendency at Hawley Bog was unimodal, peaking in late July (figure 9.5). This suggests that total system throughput and its functionality were not dependent on the presence of *H. rosa*. Because this rotifer is not an obligate associate of *S. purpurea* (Błędzki and Ellison 2003), this result was not unexpected. At Molly Bog, where *F. fletcheri* was less common and abundant (figure 9.4) and *W. smithii* is univoltine, all network metrics increased from their spring low to a midsummer high (albeit generally lower than at Hawley Bog), and remained at that level through the summer (figure 9.5).

9.4.2 A Model of Food-Web Temporal Dynamics

The occurrences of the inquilines recorded at each date were used to generate food-web models using the structure defined by Butler et al. (2008; abstracted in figure 9.1g). The abundance data were used as "storage values" to estimate fluxes between taxa (compartments) in a model based on that of Mouquet et al. (2008), but modified to include additional taxa (*M. knabi, S. gibsoni, F. fletcheri*), with bacteria (unmeasured) pooled into the detritus compartment, and measured values of plant size, volume of water, and quantity of detritus included as model inputs. Following Mouquet et al. (2008), we estimated C fluxes between compartments. All parameter values for the model are given in table 9.5. The model was run repeatedly for each set of inquilines observed in each pitcher at each sampling date.

The changes in network structure and dynamics were visualized as a series of food-web diagrams (figure 9.6). For clarity, we illustrate modeled webs at three-week intervals beginning with the second week of June and continuing through mid-September. In each network diagram, only nodes with nonzero total flows are shown. The thickness of the edges is proportional to the average flux of C and the edges are shaded from light grey to dark black, proportional to the probability of a given flow based on the frequency of its occurrence in the 40 replicates.

This visualization illustrates the strong effect of *F. fletcheri* early in the season at Hawley Bog, where its feeding on *H. rosa* dominates the flux of carbon through the system (figure 9.6). The consumption of this rotifer by *W. smithii* leads to an indirect competitive effect (dotted arrow in figure 9.6) of *F. fletcheri* on *W. smithii* that dissipates through the summer as *F. fletcheri* completes its life cycle. A similar indirect effect of *M. knabi* on *H. rosa* is also apparent, as both of these inquilines feed on POM (detritus). Similar patterns in food-web dynamics occurred at Molly Bog, although effects of *F. fletcheri* were much weaker.

Heard (1994) proposed that the *S. purpurea* food web was organized as a processing-chain commensalism. This mechanism predicts a sequence of increasing carbon fluxes from captured prey to the shredders *F. fletcheri* and *M. knabi*,

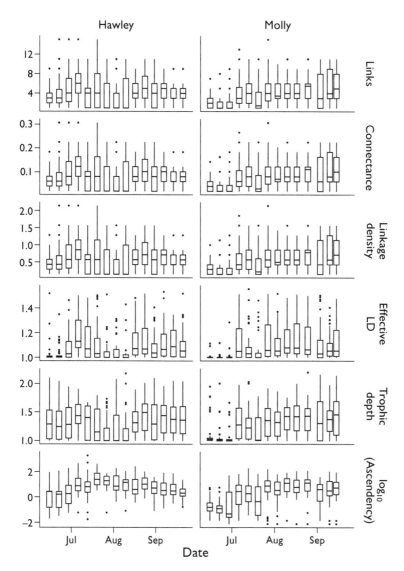

FIGURE 9.5. Time series of network and food-web metrics in *Sarracenia purpurea* food webs at Hawley and Molly Bogs. The box plots show the median, upper, and lower quartiles, and individual points in the outer deciles computed for inquiline macrobial webs sampled weekly in each of 40 pitchers. Computation of each metric is detailed in table 9.2.

TABLE 9.5. Parameter values used in our food-web model.

Variable	Description	Element	Units	Value
B	Bacteria	C	mg L^{-1}	5.00
de	Detritus	C	mg L^{-1}	371.25
N	Nitrogen	N	mg L^{-1}	0.82
Bv	Bacterivores	C	mg L^{-1}	27.85
se	Sediment	C	mg L^{-1}	0.00
wy	*Wyeomyia*	C	mg L^{-1}	303.75
α	C:N in POM	Ratio	mg mg^{-1}	6.62
θ_A	C flux from detritus	C	mg L^{-1} day^{-1}	5.39
θ_N	Flux of DIN	N	mg L^{-1} day^{-1}	0.08
m_B	Bacteria mortality	Individual	ind. day^{-1}	0.00
m_{Bv}	Bacterivore mortality	Individual	ind. day^{-1}	0.01
r_B	Bacteria respiration	C	mg day^{-1}	0.00
r_{Bv}	Bacterivore respiration	C	mg day^{-1}	0.00
r_{we}	*Wyeomyia* respiration	C	mg day^{-1}	0.01
s	Sedimentation rate	C	mg day^{-1}	0.01
u_B	Bacteria consumption of detritus	C	mg day^{-1}	0.00
u_{Bv}	Bacterivore predation	C	mg day^{-1}	0.01
u_{wy}	*Wyeomyia* predation	Individual	ind. day^{-1}	0.59
y	Plant uptake of N	N	mg day^{-1}	0.10

Modified from Mouquet et al. (2008).

through bacteria and POM, to the filter-feeding *W. smithii* (figure 9.1c). We did not find evidence for this pattern in estimated carbon fluxes, possibly because of low densities of the shredders, or high densities of *W. smithii* and *H. rosa*. As neither the presence nor the effects of *Habrotrocha* were measured either by Bradshaw (1983) or Heard (1994), its role in altering the *M. knabi–W. smithii* processing chain remains unstudied.

Two recent studies of temporal dynamics of ecological networks provide additional insights into interpreting these results. Olesen et al. (2011) studied a bipartite flower-visitation network for 12 years. Metrics of overall food-web structure (number of species, links, and connectance) were roughly constant across years, but species composition and link structure were not. Specialist butterflies and their flowers (≤2 links) were temporally more variable and had higher rates of species turnover than links between flowers and generalist visitors (>2 links). Olesen et al. (2011) suggested that phenological shifts and metapopulation dynamics drove rapid turnover of specialists, whereas redundancy in traits or functions of generalists allowed them to shift their links among other species present and "rewire" the food web. Olesen et al. (2011) concluded that global

TABLE 9.6. Coefficients used to calculate the values for the flows in our food-web model.

	an	se	de	ha	sa	me	wy	fl	re
an	0	0	$\theta_A \cdot$ Volume	0	0	0	0	0	0
se	0	0	0	0	0	0	0	0	0
de	0	$\dfrac{se \cdot de \cdot de}{\Sigma_{..}}$	0	$\dfrac{u_{Bv}(u_B \cdot ha + B) \cdot ha \cdot ha}{\Sigma_{..}}$	$\dfrac{u_{Bv}(u_B \cdot sa + B) \cdot sa \cdot sa}{\Sigma_{..}}$	$\dfrac{u_{Bv}(u_B \cdot me + B) \cdot me \cdot me}{\Sigma_{..}}$	$\dfrac{u_{Bv}(u_B \cdot wy + B) \cdot wy \cdot wy}{\Sigma_{..}}$	0	$r_B \cdot B$
ha	0	0	$m_{Bv} \cdot ha$	0	0	0	$\dfrac{u_{wy} \cdot ha \cdot wy \cdot wy}{(wy+fl)}$	$\dfrac{u_{wy} \cdot ha \cdot fl \cdot fl}{(wy+fl)}$	$r_P \cdot ha$
sa	0	0	0	0	0	0	0	0	$r_{wy} \cdot sa$
me	0	0	$m_{Bv} \cdot me$	$\dfrac{u_{Bv} \cdot me \cdot ha \cdot ha}{\Sigma_{..}}$	$u_P \cdot me \cdot sa \cdot \dfrac{sa}{\Sigma_{.}}$	0	$\dfrac{u_{Bv} \cdot me \cdot wy \cdot wy}{\Sigma_{.}}$	0	$r_{wy} \cdot me$
wy	0	0	0	0	0	0	0	$u_{wy} \cdot wy \cdot fl$	$r_{wy} \cdot wy$
fl	0	0	$m_{Bv} \cdot fl$	0	0	0	0	$u_{wy} \cdot fl \cdot fl$	$r_{wy} \cdot fl$
re	0	0	0	0	0	0	0	0	0

Source: Modified from Mouquet et al. (Mouquet et al.). The row variable equals the sum of the formulas in each cell (coefficients) times the column variable. Upper-case variable names correspond to variables in table 9.5. Lower-case variable names correspond to symbols in figure 9.6, and refer to ants (prey), sediment, detritus, *habrotrocha rosa*, *sarraceniopus gibsoni*, *metriocnemus knabi*, *wyeomyia smithii*, *fletcherimyia fletcheri*, and respiration. $\Sigma_{..}$ = *H. rosa* + *S. gibsoni* + *W. smithii* and $\Sigma_{.}$ = detritus + *H. rosa* + *S. gibsoni* + *M. knabi* +*W. smithii*.

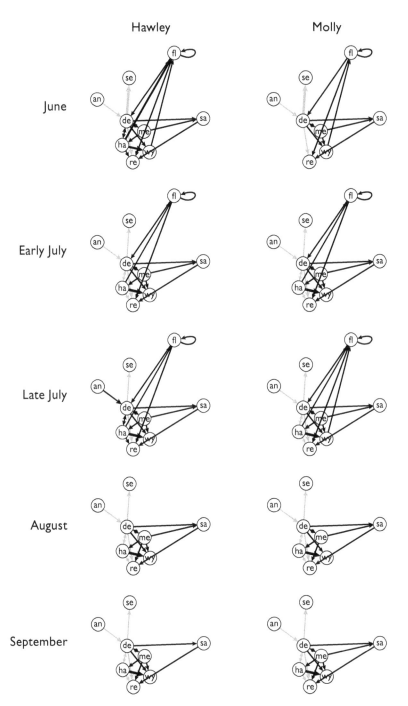

FIGURE 9.6. Temporal changes in occurrence and interaction strength in *S. purpurea* food webs at Hawley and Molly Bogs. Each web is a snapshot of the modeled *Sarracenia purpurea* assemblages at different times within a single field season. The modeled webs incorporated information on the occurrence and abundance of each taxon. The width of each arrow is proportional to estimated carbon flux between two nodes, based on the model of Mouquet et al. (2008). Model equations, parameters, and node abbreviations are given in table 9.6.

stability of network properties coexisted with local instability of network nodes and links.

Similarly, in the *S. purpurea* food web, we detected phenological and successional turnover in the obligate inquilines (figure 9.4) and parallel temporal variability in measures of network structure (figure 9.5) despite most species having >2 links. This observed global "instability" may reflect differences between multilevel and bipartite webs, or that we sampled only the more specialized, macrobial portion of the *S. purpurea* food web. If we had included the microbial generalists, the food-web structure may have been more consistent through time. However, a recent experiment by Bittleston et al. (2020) illustrates that bacterial assemblages in *S. purpurea* also have temporally variable structures dependent on initial conditions and species-specific differences in metabolism and population dynamics.

Alternatively, Cirtwill et al. (2019) identified multiple sources of variation that can contribute to the uncertainty or unreliability in constructing networks (bipartite or multitrophic) and consequently, quantifying their structure. Assuming a species pool from which the individuals making up a particular network would be drawn, these sources include variation in the probability of species interactions (as we illustrated by varying the shading of the edges in figure 9.6), variation in local occurrence (appearance and disappearance of nodes in 9.6), and variation in detection probability (implicit in our not sampling bacteria, for example).

9.5 SUMMARY

Ecological networks are not constant in space or time. As in single trophic-level ecological communities, stochastic and deterministic processes interact to produce the patterns we see in nature. The networks we observe themselves are contingent on where and when we sample them and on our ability to detect the individual species and their interactions with others species within them. The *S. purpurea* food web provides a robust observational and experimental platform with which to study network dynamics in real time within populations of pitcher plants. As we illustrate in the next chapter, unique characteristics of the *S. purpurea* food web also allow this research platform to be extended to continental spatial scales.

CHAPTER TEN

Scaling Up

The Generality of the Sarracenia *Food Web and Its Value as a Model Experimental System*

> *Et son image éveillait chez vous l'idée de l'infini; vous y constatiez, comme en un microcosme, l'existence des éléments et l'instabilité de l'univers qui se transforme à chaque minute, sous notre regard.*[1]
> —Claude Monet to Roger Marx (in Marx 1914: 291)

Ecology has few model systems and even fewer model organisms (chapter 1). Levins (1966) summarized three features of model systems—tractability, generality, and realism—that Srivastava et al. (2004) cogently argued are present in natural microcosms such as *Sarracenia* pitchers. In our two previous chapters dedicated to scaling, we illustrated these features for the plant itself. *Sarracenia purpurea* has physiological, stoichiometric, and leaf-trait scaling relationships that are comparable to those of other plants (chapter 4). It also is a tractable plant for observational and experimental studies of demography that have provided general insights into population dynamics as a function of ongoing environmental change (chapter 7). In this chapter, we explore how the *Sarracenia* microecosystem can be used as a model to answer several key questions about community-level ecological processes.

We first asked whether the *S. purpurea* food web and other inquiline webs in container habitats (Kitching 2000; Srivastava et al. 2004)—*Nepenthes* pitchers, *Heliconia* bracts, bromeliad tanks, treeholes—have similar structures. Finding that they did, we asked whether food-web structure scaled with system size. Although most authors provide little if any data on system volume, we could ask how food-web structure varies with number of taxa and their interactions. We then used data from our own work with the *S. purpurea* food web to examine two other system-independent, "scalable" questions: how does food-web structure vary with spatial extent (Buckley et al. 2003, 2010; Baiser et al. 2012), and which structural and process variables may be important to manage food webs (Gotelli and Ellison 2006a)?

10.1 THE *SARRACENIA* FOOD WEB AND OTHER CONTAINER WEBS ARE "NORMAL" FOOD WEBS

10.1.1 Food-Web Data

To place the *Sarracenia* food web in context, we used three databases. The first was the database of 358 food webs assembled in GlobalWeb[2] (Thompson et al. 2012), which includes the 213 food webs assembled by Joel Cohen into the ECOWeB database[3] (Cohen et al. 1990; Cohen 2010), the 27 predator–prey webs included in the Interaction Web DataBase,[4] and more than 100 others assembled from a variety of sources. The second was the database of 104 ecosystem webs bundled with the enaR package (Borrett and Lau 2014). The third was a database derived from the 365 tank-bromeliad food webs assembled and studied by Dézerald et al. (2013). These bromeliad webs were synthesized into 11 "composite" tank-bromeliad webs representing each of the seven species × five vegetation type × five localities sampled in French Guiana by Dézerald et al. (2013).[5] Each composite bromeliad web included any species that occurred in at least 50% of the bromeliads sampled in each species × vegetation × locality combination (*n* per combination ranged from 19 to 63). The data in the three databases came in a variety of formats. We harmonized them by deleting stray empty rows and columns, removing "nonstandard" characters, and wrangling them into square adjacency matrices (see appendix E, table E.2).

Following the convention in the enaR database, we categorized webs as either "trophic" (59 of the webs in the enaR database, all the bromeliad webs, and all those in GlobalWeb) or "biogeochemical" (45 of the enaR webs). Matrix entries in trophic webs are either zero (no trophic connection) or one (rows "eaten by" columns), whereas matrix entries in biogeochemical webs are positive real numbers representing energy fluxes from rows to columns.[6] The set of 429 trophic webs included 29 inquiline webs: 8 from 6 different *Nepenthes* species, three from *Heliconia* flower-bracts, 11 from tank bromeliads, 5 from tree-holes, and 2 from *S. purpurea* (Bradshaw 1983; Baiser et al. 2013). We augmented the dataset with the additional *S. purpurea* food web shown in figure 9.1g (giving a total of 30 inquiline and 430 trophic webs). All but two of the enaR biogeochemical webs are from "natural" (field) environments; the remaining two are webs documenting urban energy flux ("metabolism") in the cities of Beijing and Vienna.

Measures of network structure (number of nodes and links) and food-web structure (effective linkage density, effective trophic depth, and ascendency; see appendix E, table E.4) were computed for each web using the enaStructure and enaAscendency functions in the enaR package (Borrett and Lau 2014).

10.1.2 Food-Web Structure

The inquiline food webs in *S. purpurea* and other container habitats were similar in structure to those of most trophic webs (figure 10.1; table 10.1). Connectance (*C*) rarely exceeded 30%, with most taxa being connected by three to five effective links and arranged in a similar number of effective trophic levels. Connectance tended to decline with web size (i.e., the total number of taxa in a web), whereas link density (LD_E) and trophic depth (TD_E) tended to increase with web size (figure 10.2; see also Dunne et al. 2002). Likewise, network ascendency (log*A*)—the combination of total system throughput and its organization—tended to increase with network size.

Although the magnitudes of *C*, LD_E, and TD_E in the large database of food webs were consistent with previous syntheses and compilations (e.g., Cohen et al. 1990; McCann 2012), the "window of vitality"—the joint distribution of LD_E and TD_E that forms a boundary on the universe of food-web properties—was somewhat larger than expected (figure 10.3; Ulanowicz et al. 2014). We suggest that this result derived from the dependency of both LD_E and TD_E on food-web size (figures 10.2 and 10.4); many of the webs in our database were substantially larger than those used by Ulanowicz et al. (2014). This explanation was supported by the placement of the more fully resolved *S. purpurea* food web (Baiser et al. 2012) outside of the window of vitality, whereas the simpler, aggregated *S. purpurea* food web (figure 9.1g) was well within it (figures 10.3 and 10.4).

10.2 SPATIAL SCALING OF THE *SARRACENIA PURPUREA* FOOD WEB

The boundaries on link density and trophic depth described in the previous section (figures 10.3 and 10.4; see also Ulanowicz et al. 2014) suggest that a relatively small number of abiotic and biotic drivers interact to constrain food-web structure. Not surprisingly, identifying these drivers has been a major focus of research in community ecology (e.g., Pimm 1982; Paine 1988; Cohen et al. 1990; Dunne et al. 2002; Poisot et al. 2012, 2015; Mora et al. 2018). Identifying common drivers of food-web structure is challenging, however, because the elements that make up food webs—individual species—normally vary in identity at different locations. Differences in species composition of local assemblages arise because the species pools from which food webs assemble covary with expected ecological drivers— ecosystem size, resource availability, and colonization history—which themselves vary with geography and climate.

At any given location throughout its broad geographical range, however, the *S. purpurea* food web is composed of one or more subsets of the same macrobial

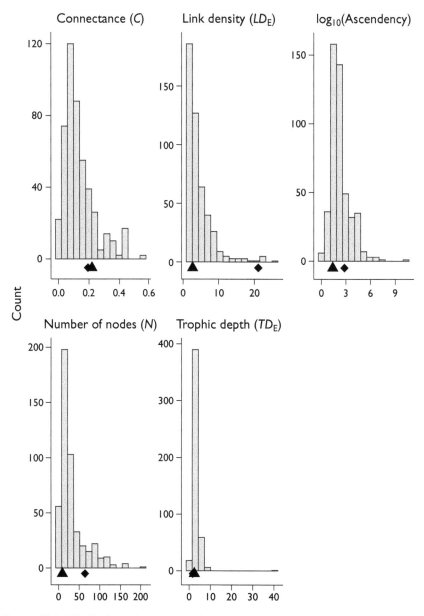

FIGURE 10.1. Distributions of network metrics (see appendix E, table E.4 for definitions and computation) across the 475 webs in our combined GlobalWeb + enaR + bromeliad webs database. The marked locations on the *x*-axes indicate the values of each metric for the 64-node *Sarracenia purpurea* web from Baiser et al. (2012) (diamond) and the 9-node summary *S. purpurea* web illustrated in 9.1g (triangle).

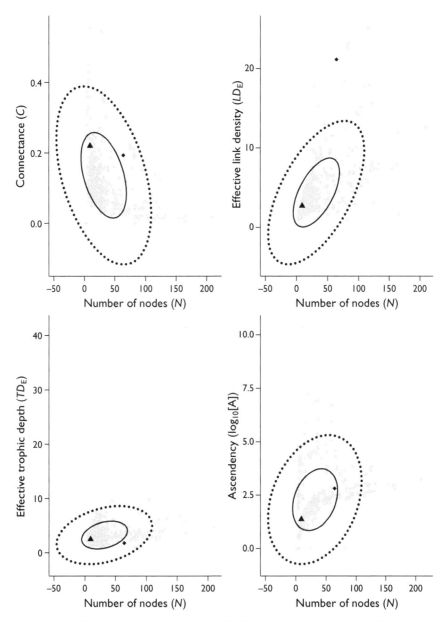

FIGURE 10.2. Size dependence of measures of food-web structure (connectance, link density, trophic depth, and ascendency; see appendix E, table E.4 for definitions and computation) on food-web size (number of nodes [species]). The points indicate the location on each biplot of each of the 475 webs in our combined GlobalWeb + enaR + bromeliad webs database. Bivariate confidence ellipses are drawn with solid (50%) and dotted (95%) lines. The diamond denotes the 64-node *Sarracenia purpurea* web from Baiser et al. (2012); the triangle the 9-node summary *S. purpurea* web illustrated in figure 9.1g; and the gray points the remaining 428 webs in the database.

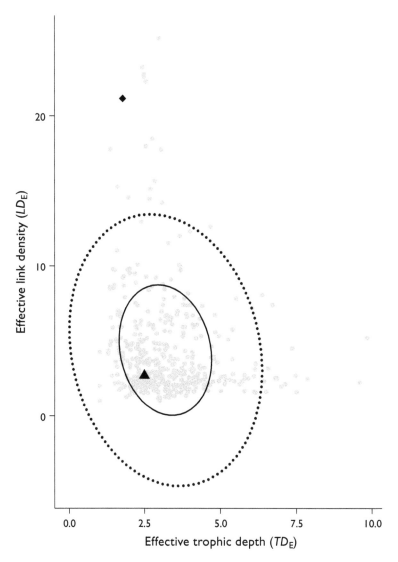

FIGURE 10.3. Joint distribution of link density and trophic depth across the 475 webs in our combined GlobalWeb + enaR + bromeliad webs database. Bivariate confidence ellipses suggesting the "window of vitality" (sensu Ulanowicz et al. 2014) are drawn with solid (50%) and dotted (95%) lines. The locations in the window of vitality of the 64-node *Sarracenia purpurea* web from Baiser et al. (2012) and the 9-node summary *S. purpurea* web illustrated in 9.1g are indicated by the black diamond and triangle, respectively.

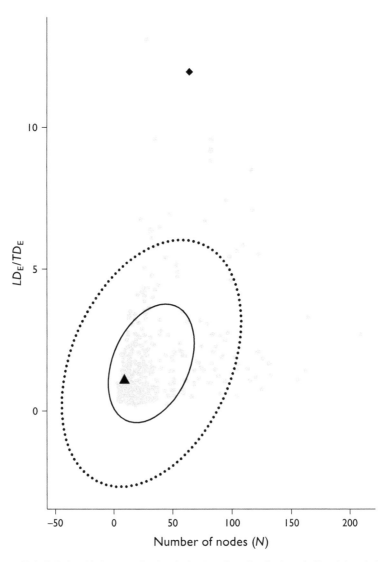

FIGURE 10.4. Relationship between food-web size (number of nodes [species]) and the window of vitality (here illustrated as the ratio between link density and trophic depth: LD_E/TD_E) across the 430 webs in our combined GlobalWeb + enaR + bromeliad webs database. Bivariate confidence ellipses are drawn with solid (50%) and dotted (95%) lines. The locations on the plot of the 64-node *Sarracenia purpurea* web from Baiser et al. (2012) and the 9-node summary *S. purpurea* web illustrated in figure 9.1g are indicated by the black diamond and triangle, respectively.

TABLE 10.1. Five metrics of food-web and network structure for all 430 trophic webs (including the 30 inquiline webs), the different kinds of inquiline webs, 43 "natural" biogeochemical webs, and two urban-metabolism webs in our combined GlobalWeb + enaR + bromeliad web database.

	N	C	LD_E	TD_E	$\log(A)$
All trophic webs	22.0	0.1	3.5	3.0	2.0
	33.8 (32.19)	0.1 (0.08)	4.6 (3.79)	3.3 (2.28)	2.1 (1.07)
Inquiline webs					
S. purpurea	9.0	0.2	2.7	2.2	1.4
	26.3 (32.65)	0.1 (0.04)	8.5 (11.01)	2.2 (0.36)	1.7 (1.06)
Nepenthes spp.	12.5	0.1	2.2	3.2	1.4
	12.8 (4.89)	0.1 (0.03)	2.3 (0.77)	3.0 (0.84)	1.3 (0.41)
Heliconia spp.	7.0	0.1	1.9	1.7	0.7
	7.7 (1.15)	0.1 (0.02)	1.9 (0.11)	1.7 (0.09)	0.7 (0.12)
Bromeliads	13.0	0.2	4.2	1.8	1.4
	15.9 (5.30)	0.2 (0.06)	4.4 (0.43)	2.1 (0.54)	1.6 (0.27)
Treeholes	12.0	0.1	2.0	2.4	1.3
	11.2 (1.3)	0.1 (0.06)	2.2 (0.78)	2.6 (0.46)	1.2 (0.31)
Biogeochemical webs					
"Natural" webs	7.0	0.3	2.1	2.9	4.0
	13.0 (14.73)	0.3 (0.13)	2.1 (0.49)	2.9 (0.89)	4.0 (1.69)
Urban metabolism	8.0	0.5	2.7	2.6	3.3
	8.0 (1.41)	0.5 (0.10)	2.7 (0.36)	2.6 (0.01)	3.3 (1.08)

For each class of webs, medians are given in the first row, and means (SD) are given in the second row. See appendix E, table E.4 for definitions and computations of the different metrics.

taxa and the majority of its microbial taxa (figure A.8; Buckley et al. 2003, 2010). This common species pool (a continental-scale "meta-web" sensu Poisot et al. 2012) is, as far as we know, unique among meta-webs and their subsetted local webs ("local realizations" sensu Poisot et al. 2012). The common meta-web of *S. purpurea* has allowed us to assess natural relationships between food-web structure of its inquilines and geography (latitude, longitude), climate (precipitation, temperature), N inputs (see also chapter 3), habitat size (pitcher volume), and the presence of the top predator (*Wyeomyia smithii*) without the complication of species pools that themselves covary with these drivers. We looked at these relationships using 780 replicate food webs sampled from *S. purpurea* pitchers at 39 sites ranging from northern Florida to Newfoundland in eastern North America and across Canada to eastern British Columbia (figure A.8; Baiser et al. 2012).

Among individual pitchers within sites, LD_E and TD_E of local webs were positively correlated. Both LD_E and TD_E also increased with local species number, habitat size (i.e., pitcher volume), and latitude (figure 10.5); latitude itself was strongly correlated with climatic variables related to *S. purpurea* food-web structure (Baiser et al. 2012). However, even though pitcher volume and latitude had the strongest associations with the number of species in the local webs, these predictor variables never explained more than 5% of the observed variance in LD_E, TD_E, or other food-web metrics.

Instead, abiotic and biotic variables were better predictors of *S. purpurea* food-web structure among sites. The abundance of *W. smithii* decreased with latitude across sites (i.e., replicate sampled pitchers pooled within sites), and the number of species in local webs concomitantly increased (figure 10.6a; Buckley et al. 2003, 2010). This effect was significant for both bacteria and protozoa (figure 10.6b), but not for macroinvertebrates (Buckley et al. 2003). We hypothesized that this reverse latitudinal gradient in species richness resulted from the release of microbes from predation pressure by *W. smithii* (Buckley et al. 2003). In a subsequent analysis, 33% in the variation in trophic depth of *S. purpurea* food webs could be explained by a combination of climatic variables (captured by latitude) and *Wyeomyia* abundance (Baiser et al. 2012).

10.3 THE *SARRACENIA PURPUREA* FOOD WEB AS A MODEL EXPERIMENTAL SYSTEM FOR UNDERSTANDING AND MANAGING FOOD WEBS

Habitat size (e.g., available area, patch size, pitcher size, or tank volume), predation pressure, resource availability (prey or detritus), dispersal, and colonization all interact to structure food webs in marine, freshwater, and terrestrial habitats (Paine 1966; Pimm 1982; Cohen et al. 1990; Thompson and Townsend 2005; Östman et al. 2007; McCann 2012; Petermann et al. 2015a,b; Romero et al. 2016). The roles of each of these common drivers and the importance of interactions among them have been determined most frequently through compilations of large datasets (e.g., Cohen et al. 1990; Borrett and Lau 2014) and models (e.g., Morton et al. 1996; Jordán et al. 2012; Mora et al. 2018; Weterings et al. 2018; Cirtwill et al. 2019). Replicated experiments have been less common because of the large size of the habitat required to support a food web and because of spatial variability in meta-webs and their local realizations (Poisot et al. 2012).

We and others have taken advantage of the small habitat size and representative structural complexity (figures 10.1–10.4 and table 10.1) of the *S. purpurea* food web to conduct well-replicated experiments on mechanistic causes of its structure (chapter 9, §9.1.2). These experiments have taken on additional urgency as anthropogenic modifications of the global environment are changing species composition (Dornelas et al. 2019), reducing population sizes, and accelerating

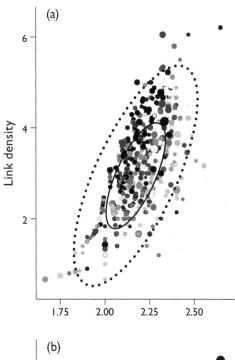

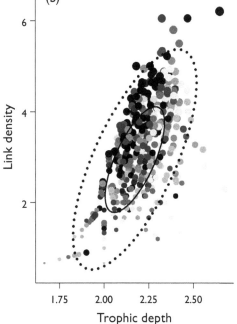

FIGURE 10.5. Joint distributions of link density and trophic depth across the 780 replicate *Sarracenia purpurea* food webs we sampled at 39 sites from Florida to British Columbia. In both panels, the shading of the points darkens with increasing latitude. In (**a**), the size of the points is proportional to the volume of the pitcher, whereas in (**b**), the size of the points is proportional to the number of species in each local web. Bivariate confidence ellipses suggesting the window of vitality (sensu Ulanowicz et al. 2014, and see figure 10.3) are drawn with solid (50%) and dotted (95%) lines.

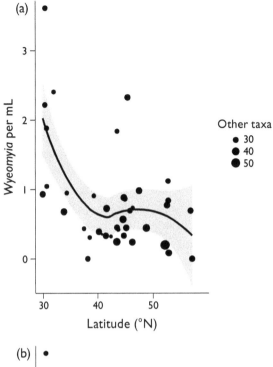

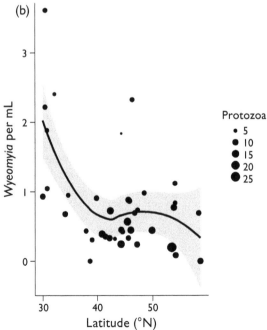

Figure 10.6. As abundance of larvae of the mosquito *Wyeomyia smithii* declined with latitude, the total number of other taxa (**a**) or protozoa alone (**b**) present in the *S. purpurea* food web increased.

species extinctions at local and global scales (Leakey and Lewin 1996). These changes threaten the integrity and functioning of food webs and entire ecosystems (e.g., Lotze and Milewski 2004; Ebenman and Jonsson 2005; Dunne and Williams 2009; Yeakel et al. 2014; Valiente-Banuet et al. 2015; Ellison 2019; Pringle et al. 2019).

10.3.1 Fishing Down the Sarracenia Food Web

The experimental tractability of the *S. purpurea* microecosystem enabled us to do the first experiment that tested how changes in both trophic structure and habitat size could determine the relative abundance of other species in a complete food web (Gotelli and Ellison 2006a). In short, we asked whether removing the large predatory species in an aquatic food web had similar effects on the *S. purpurea* food web as harvesting out the largest fish in oceans, rivers, and lakes has had on marine and freshwater food webs (Pauly et al. 1998, 2000).

Working in a large peatland in northeastern Vermont (Moose Bog), we located 50 adult *S. purpurea* plants and assigned them randomly to one of five treatment groups: (1) **trophic removal**, in which we removed the pitcher water and the larvae of the dipterans *Wyeomia smithii*, *Fletcherimyia fletcheri*, and *Metriocnemus knabi*; counted the larvae; and refilled the pitcher with an identical volume of distilled water (dH_2O); (2) **habitat expansion**, in which we censused the dipteran larvae, returned them to the pitcher, and then filled the pitcher to the brim with additional dH_2O; (3) **trophic removal + habitat expansion** (an additive combination of the previous two treatments); (4) **trophic removal + habitat contraction**, in which we removed both the pitcher fluid and the dipteran larvae; and (5) **controls**, in which we removed the fluid and the dipteran larvae and then immediately replaced them.

These treatments not only mimicked changes in habitat volume (expansion or contraction) and effects of top-down control (removal or retention of dipteran larvae), but also changed bottom-up effects because prey capture increased with habitat volume. Prey capture was highest in the habitat expansion treatments, intermediate in the controls, and lowest in the habitat contraction treatment (see also Wolfe 1981; Cresswell 1993). Each treatment was applied to all of the pitchers on every plant in the treatment group (Gotelli and Ellison 2006a).

We nondestructively censused the macrobes and protozoa in the experimental plants weekly throughout the growing season in the summer of 2000. As a response variable, we calculated the average abundance through time of each macrobial taxon (the three dipteran larvae, *Sarraceniopus gibsoni*, and *Habrotrocha rosa*) and the abundance of all protozoa pooled within each leaf of each treatment. We then used path analysis (Wootton 1994) to compare the fit of these abundance data to different food-web models: single-factor models, in which abundances of each species were hypothesized to respond individually to differences in habitat volume, prey, or a single predator; and food-web models, in which abundances of each species were hypothesized to respond to direct and indirect trophic interactions (i.e., the network paths of figure 9.1g). For the food-web models, we included in the path analysis a latent variable to represent the unmeasured effects of the uncensused bacterial assemblage within each pitcher (Gotelli and Ellison 2006a).

The best-fitting model was the single-factor predation model of top-down control of abundances by *W. smithii* (Gotelli and Ellison 2006a). More complex food-web models fit the data only marginally less well than the top-down predation model. Inclusion of bacteria as a latent variable did improve the fit of the food-web models, suggesting a complex interaction of bottom-up and top-down control (e.g., the processing-chain commensalism described by Heard 1994). Neither the fit of the food-web models (with or without bacteria) nor the fit of the single-factor predation model improved with the inclusion of habitat volume in the path analysis.

10.3.2 Is Wyeomyia smithii *a Keystone Predator?*

Our experiment modifying *S. purpurea* trophic structure and habitat size illustrated that predator abundance alone and in concert with other species in the food web could alter abundances of all other species in the network, and that these effects can be much stronger than those of habitat size (Gotelli and Ellison 2006a). Top-down control is a major driver of food-web structure across ecosystems (Menge 2000; Shurin et al. 2006; Frank et al. 2007; Gruner et al. 2008). Experiments in the past 15 years have identified linkages between trophic structure and habitat size in aquatic container habitats (e.g., Srivastava 2006; Petermann et al. 2015a,b) and terrestrial islands (e.g., Pringle et al. 2019), buttressing Pauly and colleagues' (1998; 2000) conclusion that removal of top predators has strong effects on trophic depth and the abundance of prey species.

Similarly, our single-factor *W. smithii*-predation model correctly identified the role of its larvae as predators of rotifers (see also Addicott 1974; Błędzki and Ellison 1998) and as prey of larger *F. fletcheri* larvae (see also Butler et al. 2008). Removal of *W. smithii* led to increased abundance of *H. rosa, S. gibsoni*, and protozoa, each of which operate at different trophic levels (figure 9.1g). Collectively, the results of the path analyses were consistent with other studies that found top-down and bottom-up effects of *W. smithii* on the *S. purpurea* food web (Gotelli and Ellison 2006a). But these results do not mean that *W. smithii* was acting as a keystone predator.

As originally conceived by Paine (1966), keystone predators increase the diversity of taxa at lower trophic levels by preferentially preying on species that are competitive dominants (Valls et al. 2015). Subordinate species are then freed from competitive exclusion and increase in abundance. Gotelli and Ellison (2006a) did not explicitly test for competitive release of inquilines feeding at lower trophic levels. However, the presence of *W. smithii* decreased the abundances of all prey species that we measured. By this criterion, *W. smithii* exerted strong top-down control on the *S. purpurea* food web, but it did not increase species richness of competitively subordinate prey as a classic keystone predator does.

Our results mirrored those of Addicott (1974), who deliberately set out to test whether *W. smithii* was a keystone predator. Addicott (1974) also found that *W. smithii* preyed indiscriminantly on rotifers and protozoa, leading to a progressive loss of rare species with increasing predation pressure. However, a reason that neither our experiments nor Addicott's (1974) detected keystone effects of *W. smithii* may have been because that we did not quantify diversity or abundance of bacteria in the food webs that we studied. When bacterial diversity has been examined explicitly, classical keystone effects of *W. smithii* always have been found (Cochran-Stafira and von Ende 1998; Kneitel and Miller 2002; Peterson et al. 2008). This dependency of experimental results on taxonomic resolution reinforces the idea that taxonomic lumping and coarse resolution of food webs may hide important biological details (Paine 1988; Polis 1991).

10.3.3 Dynamic Food Webs in Dynamic Habitats

In the mid-1990s, when we began our work with *S. purpurea*, food-web studies neglected spatial or temporal dynamics, and metacommunity ecology was in its infancy. In recent years, spatiotemporal dynamics have been studied, documented, and modeled in food webs and other networks (e.g., Poisot et al. 2015; Pilosof et al. 2017; Piovia-Scott et al. 2017; Hutchinson et al. 2019; Schiaffino et al. 2019) and in metacommunities (e.g., Ricklefs 2008; Fernandes et al. 2013; Datry et al. 2016; Harvey et al. 2020). But how complete food webs (*not* single trophic-level metacommunities) assemble in a habitat that itself is changing through time and that interacts with the food web remains unstudied. We do not know whether this lack of attention results from the rarity of such situations, our failure to perceive them, or the difficulty in studying them. However, we do know that living container habitats such as bromeliads, *Heliconia* bracts, and pitcher plants (table 10.1) provide opportunities to study experimentally the dynamics of a food web that reassembles at predictable intervals in a habitat that itself is a living, growing organism.

TABLE 10.2. Binary values used to calculate saturation of the *S. purpurea* macrobial food web, and their decimal equivalents.

Taxon	Binary Value	Decimal Value
Metriocnemus knabi	1	1
Habrotrocha rosa	10	2
Sarraceniopus gibsoni	100	4
Wyeomyia smithii	1000	8
Fletcherimyia fletcheri	10000	16

The saturation (or completeness) of each of the 32 possible food webs can be quantified by assigning a binary value to the presence or absence of each taxon and summing the decimal values of their binary representations.

Our experimental work on food webs in dynamic habitats was built on two previous studies. First, Bradshaw and Creelman (1984) experimentally showed that *S. purpurea* and its inquilines are mutualists. Processing of captured prey by inquilines releases CO_2 and NH_4 that are taken up by the plant (Joel and Gepstein 1985). The plant, in turn, oxygenates the pitcher fluid, preventing a switch to anaerobic conditions, which can nonetheless occur if prey are superabundant (see also chapters 12 and 13). Second, following Fish and Hall (1978), we identified clear temporal patterns in inquiline colonization, development, and maturation or eclosion in the inquilines (figures 9.4–9.6) that we could experimentally manipulate (Ellison et al. 2003).

Working at Swift River Bog, a 1.9-ha glacial kettle bog in central Massachusetts, we conducted three experiments to examine the joint effects of environmental variability and the plant itself on the structure of the macrobial inquiline food web (Ellison et al. 2021). In all three experiments, we quantified food-web structure with a saturation index that ranged from 0 (no species present) to 31 (all 5 macrobe species present). Higher values of saturation indicate more complete webs, with greater numerical weight being placed on the presence of higher trophic levels (*F. fletcheri* and *W. smithii*) (table 10.2).

Our first two experiments examined the impacts of atmospheric deposition of N and P (chapter 3) on food-web assembly and structure. One of these experiments was a "pulse" experiment (sensu Bender et al. 1984) in which nutrients were added once, early in the life of each pitcher, and effects on the inquilines subsequently were assessed. The other was a "press" experiment (sensu Bender et al. 1984) in which nutrients were added every 2 weeks throughout the growing season. Both experiments were factorial response-surface designs (Cottingham et al. 2005) in which five levels each of NH_4NO_3 (4.7, 9.4, 18.8, 37.5, and 75.0 mM N) and PO_4 (0.63, 1.3, 2.5, 5.0, and 10.0 mM P), or dH_2O as a control were added to pitchers.

Pulse Experiment   In the first (pulse) experiment, one of the 26 nutrient levels was added to every pitcher of 130 nonflowering plants that were spatially grouped in five replicated "blocks" of 26 plants each. Each plant within a block was randomly assigned to a nutrient treatment (one of the 25 possible N × P combinations or the dH₂O control); 20 mL of nutrient solution was added once to each pitcher 2–3 days after it had opened and the leaf had fully hardened. The first leaves were treated on June 4, 2001, and plants were checked every 3 days for new pitcher production. Macrobial inquilines were destructively censused in every treated leaf of a single block at 3-week intervals (June 25, July 16, August 6, August 27, and September 17).

In this pulse experiment, food-web saturation peaked at intermediate levels of N added (figure 10.7) and was highest in midsummer in leaves produced earlier in the season ($P = 0.03$). Although the amount of P added did not significantly alter food-web saturation ($P = 0.42$), there was a significant N × P interaction ($P = 0.04$; figure 10.7).

Press Experiment   In the press experiment, which was conducted concurrently with the pulse experiment, we only had one block of 26 plants and one replicate plant per N × P (or dH₂O) combination. Food webs were sampled nondestructively every 2 weeks, at which time additional nutrient solution was added to maintain the total pitcher-fluid volume of 20 mL.

As in the pulse experiment, food-web saturation peaked at intermediate levels of N added (figure 10.7) and was highest in midsummer ($P < 0.001$) in leaves produced earlier in the season ($P = 0.04$). Similar to the pulse experiment, the amount of P added did not alter food web saturation ($P = 0.5$), but in contrast to the pulse experiment, there was no significant N × P interaction ($P = 0.6$; figure 10.7).

Experimentally Testing Interactions between S. purpurea *and Its Food Web*   In the third experiment, done 3 years after the pulse and press experiments, we quantified the joint effects of the living plant, nutrient additions, prey additions, and predation pressure on the structure of the *S. purpurea* macrobial food web. This experiment was a four-treatment factorial/response-surface experiment that used 212 adult *S. purpurea* assigned randomly to unique combinations of treatment levels.

The first treatment tested the influence of the living plant itself on the food web. Pitcher interiors received one of three treatments: (1) pitchers were fitted with intact polyethylene tubes (2.2-cm diameter) that physically isolated the imquilines from the pitcher fluid and from interactions with the pitcher ($n = 64$ plants with polyethylene tubes); (2) pitchers were fitted with polyethylene tubes that had holes punched out of them to allow for fluid exchange and interactions between the inquilines and the plant ($n = 64$ plants with polyethylene controls); (3) pitchers

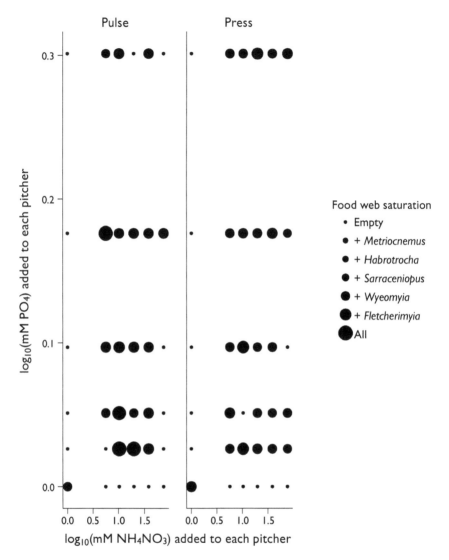

FIGURE 10.7. Food-web saturation in pitchers fed one of 25 combinations of nitrogen (N) and phosphorus (P) or distilled water once, 3 days after the pitcher opened (the "pulse" experiment; left panel); or 3 days after the pitcher opened and every 2 weeks thereafter (the "press" experiment; right panel). This figure illustrates data from the first leaf of the season (fed on or beginning the week of June 4) at the mid-summer harvest (August 6 for the pulse experiment, July 23 for the press experiment), when saturation was highest.

were established as unmanipulated controls without polyethylene tubes ($n = 64$ plants with no manipulation).

The second treatment examined the effects of N deposition (environmental variability). One of eight different concentrations of NH_4NO_3 (0, 1.6, 3.1, 6.3, 12.5, 25, 50, or 100 mg N/L) was added to a randomly selected plant; there were $n = 8$ replicates of each N concentration in each set of 64 polyethylene-tube, no-tube, or tube-control treatments. Because there was constant uptake of NH_4NO_3 in both no-tube and tube-controls, we supplied an equivalent $[NH_4NO_3]$ to each of these plants each week. Plant liquid volume was maintained at two-thirds full by adding nutrient solutions or dH_2O water as necessary. Plants with intact tubes were not able to assimilate NH_4NO_3 supplied to the tubes. We assumed loss due to volatilization and bacterial use was small relative to plant assimilation and therefore we simply topped off each tube to two-thirds full every week.

The third treatment examined effects of prey availability (organic nutrients). Supplemental prey (10 lab-reared *Drosophila melanogaster*/week) were added to one-half of each inorganic N × tube combination.

The fourth treatment assessed the influence of the top predator on the development of the macrobial food web. Second-instar *F. fletcheri* either were removed from pitchers (one-half of each N × tube × fruit fly treatment combination) or added to pitchers (one-half of each N × tube × fruit fly treatment combination).

In total, this response-surface design resulted in $n = 2$ plants in each tube × N × fruit fly × *F. fletcheri* combination. Finally, we also marked 20 randomly chosen additional plants as untouched, unmanipulated controls.

One treated plant from each treatment combination (96 pitchers) and half the unmanipulated controls were harvested after 4 weeks and the remainder after 8 weeks. Food-web saturation was quantified as in the pulse and press experiments.

Isolating the macrobial food web from interactions with the plant had a strong effect on food-web saturation. Food webs in pitchers without tubes or with perforated tubes were significantly more saturated than food webs in pitchers with tubes ($P = 0.0003$; figure 10.8). Food webs in intact tubes rarely had predators (*W. smithii* or *F. fletcheri*), because adults of these Diptera did not oviposit in intact tubes. Removal of *F. fletcheri* from plants led to an increase in food-web saturation ($P = 0.0003$; figure 10.8, left panel), probably because this predator consumed both rotifers and *W. smithii* larvae. Additions of prey or inorganic nitrogen alone did not alter food-web structure ($P = 0.4$, both cases). However, there was a nutrient × *F. fletcheri* interaction ($P = 0.03$): in the presence of *F. fletcheri*, increasing NH_4NO_3 led to decreased saturation of the food web. In contrast, in the absence of *F. fletcheri*, increasing NH_4NO_3 led to increased saturation of the food web. None of the other two-way, three-way, or four-way interaction terms were significant ($P > 0.2$, all cases).

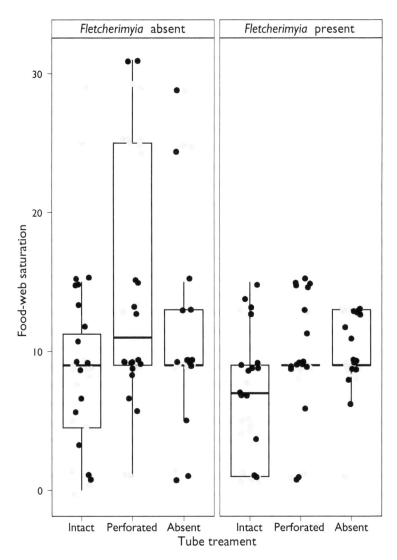

FIGURE 10.8. Effects of tube treatment manipulation of the top predator on food-web saturation in the *Sarracenia purpurea* macrobial food webs in the four-level tube × inorganic nitrogen × fruit-fly × *Fletcherimyia* manipulation. Data are pooled among nutrient and addition treatments, for which the main effects were not statistically significant. In each panel, the effects of the three levels of the inserted tubes (intact, perforated, absent) are ordered along the *x*-axis; the two different panels illustrate effects of the tube treatment without (left) or with (right) the top predator *F. fletcheri* present in the pitcher. Box-and-whisker plots illustrate median (dark horizontal line), upper and lower quartiles (top and bottom of the box), and upper and lower hinges (length of vertical "whiskers" = 1.5 × interquartile range). Individual points represent data from each pitcher harvested in mid-summer (grey points) or at the end of the growing season (black points).

10.4 SYNTHESIS

The food webs of inquilines that assemble in the growing and developing pitchers of *S. purpurea* have topological structures that are very similar to those documented for food webs in any other habitat. Likewise, bottom-up and top-down processes and keystone predators control dynamics in the *S. purpurea* food web, just as they do in widely studied marine, freshwater, and terrestrial food webs. The species that make up the food webs within individual pitchers at any site are drawn from a meta-web, whose core members are obligate mutualists that co-occur with the plant across the North American range of *S. purpurea*. This uniquely geographically large and consistent meta-web combined with the ease of doing manipulative experiments with *S. purpurea* and its inquilines, has enabled ecologists to test models and theories of food-web dynamics independent of geographical and climatic turnover in species composition. Finally, the habitat in which the *S. purpurea* food web assembles—the plant itself—is not an inert container or an external "environment" but rather is a codeterminant of dynamics in this model ecological food web.

Part IV

Tempests in Teapots

CHAPTER ELEVEN

Context

Tipping Points and Regime Shifts

*[T]out ceci est ce qu'il y a de mieux; car s'il y a un volcan
à Lisbonne, il ne pouvait être ailleurs; car il est impossible
que les choses ne soient pas oùelles sont, car tout est bien.*[1]
—Voltaire (1759:36)

The concepts of tipping points and regime shifts have become popular both with ecologists, who are grappling with environmental change, and with the public at large, who are interested in social and economic trends. But the underlying mathematical models often seem difficult to relate to these phenomena, and it is difficult to even define what constitutes a regime change or an alternative stable state.

"Tipping points" and "regime change" are mainstays in popular culture (Gladwell 2000) and buzzwords in scientific discourse. Nearly 325,000 books, papers, conference proceedings, etc. on these topics have published since 1957 (when, as discussed in appendix F, "tip point" entered the lexicon), more than 16,000 of which are in the environmental sciences sensu lato.[2]

Ecologists are not immune to these fashions: "regimes" and the "tipping points" between them have proliferated in the literature and have been loosely applied to a variety of phenomena. For example, either side of an ecotone, edge, or boundary with "stable structures, functions, processes, and feedbacks" (Angeler et al. 2016) has been characterized as a "regime." Ecological changes from the small and continuous (e.g., changes in hydrologic flow, frequency of fire and other "disturbances," or fish harvesting rate: Poff et al. 1997; Freeman et al. 2017; Harvey et al. 2017; Thom et al. 2017; Goetze et al. 2018) to the large and abrupt (e.g., thawing of permafrost, entire turnover of fish populations and fisheries, or different phases of the Pacific Decadal Oscillation: Mantua and Hare 2002; Fogarty et al. 2016; Schuur and Mack 2018) are regularly characterized as regime shifts, regime changes, or tipping points. Some have even suggested that the entire planet Earth is near, passing, or just past a planetary tipping point (Rockström et al. 2009; Barnosky et al. 2011, 2012, 2013, 2016; Steffen et al. 2015, 2018; but see Brook et al. 2013; 2018)

This chapter provides background, examples, and criteria we need for identifying ecological tipping points, regime shifts. We also address the potential use of advance warning of tipping points in forestalling or managing undesirable regime shifts. The associated appendix (appendix F) places tipping points and regime shifts in broader sociocultural and political contexts, and also describes the types of general observations and experiment that would give evidence for any sort of tipping point or regime shift, and the implied alternative stable states. Chapter 12 describes alternative states in the *Sarracenia* microecosystem and proposes it as an ideal model system for studying the widespread problem of nutrient enrichment and the eutrophic collapse of aquatic ecosystems. Finally, chapter 13 "scales down" the processes of tipping points and regime shifts to a mechanistic level by examining the metaproteomic response of the *Sarracenia* microecosystem to enrichment. The molecular diversity of the metaproteome provides a richer and more satisfying measure of "ecosystem function" than traditional metrics such as biomass accumulation, decomposition rate, or $[O_2]$. Intriguingly, higher-level measures of ecosystem function (here $[O_2]$) recover but lower-level proteomic profiles do not. This result calls into question whether these systems have really returned to the same initial state and suggests further directions for research in tipping points and regime shifts.

11.1 BACKGROUND

Regime shifts most often are induced by slow changes in one or more external driver variables, such as atmospheric $[CO_2]$ or $[P]$ in a lake. Slow increases in these environmental drivers nevertheless can cause rapid changes in response variables such as air temperature or the biomass of cyanobacteria in a lake (Carpenter and Brock 2006; Ratajczak et al. 2018). Regime shifts are identified both by an abrupt change in slope of a time series of observations of such response variables (e.g., May 1977; Zeileis et al. 2003; Yasuhara et al. 2008) and by a change in the causal relationship between the response variable(s) and the underlying driver (e.g., Yasuhara et al. 2008; Bestelmeyer et al. 2011; Ratajczak et al. 2018). Figure 11.1 illustrates how a slowly increasing driver can be associated with a discrete step-change in the level of a response variable. A characteristic of many regime shifts is that, even after the driver returns to low levels, the system exhibits "hysteresis": the response variable remains in the new state for prolonged periods of time because of stabilizing feedback loops in the system (appendix F, figure F.5; With and Crist 1995; Carpenter 2003; Pascual and Guichard 2005; Litzow et al. 2008; Odion et al. 2010; Petraitis 2013).

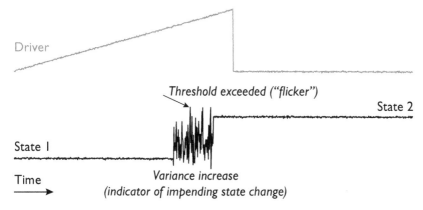

Driver

Threshold exceeded ("flicker")

State 2

State 1

Time

Variance increase
(indicator of impending state change)

FIGURE 11.1. General driver–response–feedback model of regime shifts. Gradual changes in drivers begin to alter response variables. The variance in the response increases prior to the regime shift (Carpenter and Brock 2006; Dakos et al. 2008) and the system may even "flicker" above a threshold well before the regime shift is consistently evident (Brock and Carpenter 2010). In either state, negative feedbacks and hysteresis in the system may slow or prevent the system changing from one regime to another (Beisner et al. 2003; Laio et al. 2008).

11.1.1 Examples of Regime Shifts and Alternative States

Some of the clearest examples of ecological regime shifts have been found in freshwater and marine ecosystems. For example, Carpenter and his colleagues (Carpenter et al. 1992, 2001; Carpenter and Kitchell 1993; Carpenter 2003; Ludwig et al. 2003) have shown that long-term, slow increases in phosphorus loading of lakes can cause rapid shifts from oligotrophic to eutrophic states. Freshwater food webs can show either abrupt transitions resulting from time available for growth of predators and prey (Steiner et al. 2009) or slow transitions following the reintroduction of top predators (Mittelbach et al. 2006; Houseman et al. 2008). Similarly, gradual changes in oceanographic circulation have been associated with rapid changes in the structure of marine food webs and dramatic shifts in fish harvests (Rodionov 2005b). Alternative states—the starting and ending points of regime shifts (Schröder et al. 2005)—have been induced experimentally in the lab in aquatic microcosms (Steiner et al. 2009), and in the field in the rocky intertidal (Petraitis and Latham 1999; Dudgeon and Petraitis 2005; Petraitis et al. 2009), lakes (Carpenter et al. 2001; Scheffer and Carpenter 2003; Mittelbach et al. 2006; Houseman et al. 2008), pitcher plants (chapter 12; Sirota et al. 2013), and high-elevation snowfields (Sturm et al. 2005).

11.1.2 Linking Empirical Data with Mathematical Models of Alternative States

Empirical analyses of alternative states involve repeated measurements of drivers and the "response" variables of the system that they affect to demonstrate the passage of the system from one state to another. However, there is little agreement among empiricists as to which response variables or criteria should be used to recognize different states, and what kinds of data and experiments are needed to demonstrate that these states are stable (see appendix F, §F.4).

We suggest that support for the existence of alternative states should come in the form of comonitored temporal trajectories of driver variables and response variables (Bestelmeyer et al. 2011). In either a controlled field or lab experiment or an observational field study, we need a long time series of repeated measures of both variables.[3] Although ecologists typically use widely spaced census intervals to ensure statistical independence of their samples (Dornelas et al. 2013), much more frequent sampling is required to reliably document state changes because the autocorrelation function is an important part of the signature of alternative states (Carpenter and Brock 2006; Dakos et al. 2012). Unfortunately, this kind of monitoring (long time series with frequent, autocorrelated measurements) is rare, and it is doubtful that monitoring of traditional indicator variables can provide enough lead time for intervention to prevent a regime shift (figure 11.2; Biggs et al. 2009; Contamin and Ellison 2009).

Further, we should be able to link empirical observations to a mathematical model. Observations of multimodality, nonlinearity, critical slowing down, and hysteresis (appendix F, §F.4) need to be linked to an explicit mathematical model that describes changes in these variables and predicts true bi- or multiphasic behavior. Ideally, the parameters for such a model should be estimated independently for the same system, and sensitivity of the model predictions explored over a systematic coverage of realistic parts of the parameter space (Sirota et al. 2013; Lau et al. 2018). The general model developed by Scheffer and Carpenter (2003, see also Scheffer et al. 2009) is a good starting point.

Finally, we should be able to reproduce observed "natural" patterns or model predictions in controlled, spatially replicated experiments (Boettiger and Hastings 2012). But even without manipulation and replication, it remains challenging to gather the long time series of drivers and state variables needed for robust analysis (Bestelmeyer et al. 2011). We also need a null hypothesis to contrast with a pattern generated by alternative states. Because most ecological systems are dynamic, a stable state is neither a pattern of stasis nor one of simply no relationship between a driver variable and a state variable. Rather, we suggest that the simple model of environmental tracking (appendix F, figure F.3 top) is an appropriate null hypothesis. If the state variable tracks the driver, then we may still see changes between states that could appear nonlinear (Ratajczak et al. 2018). There might even be time lags present between a change in the driver and a change in the response variable. However, those time lags should be relatively constant and not differ for high and low levels of the driver.

11.2 A POTENTIAL NEED FOR INTERVENTIONS

Recent dramatic regime shifts in ecological, biophysical, and economic systems also have highlighted the need for better advance warning of regime shifts (e.g., Dakos et al. 2008; Scheffer 2009; Scheffer et al. 2009; Pace et al. 2015; Dakos et al. 2019). However, it remains an open question whether indicators that presage regime shifts also can be used to manage, mitigate, or forestall them altogether (Biggs et al. 2009; Contamin and Ellison 2009; Suding and Hobbs 2009; Pace et al. 2015).

Biggs et al. (2009) and Contamin and Ellison (2009) showed that preventing an undesirable regime shift requires far more advance warning—by many decades— than current indicators used by ecologists and policymakers provide (figure 11.2). Contamin and Ellison (2009) analyzed Carpenter and Brock's (2006) model of regime shifts in lakes and suggested that under a "business-as-usual" scenario, nearly 50 years of advance warning of a tipping point would be required to reduce below 50% the probability of a regime shift from a clear, oligotrophic lake to a murky, eutrophic one. The ability to respond more rapidly or more aggressively to leading indicators reduced the lead time required (right side of figure 11.2) to avert a regime shift. Comparable analyses of fisheries models by Biggs et al. (2009) gave similar results. Our work with the *Sarracenia* microecosystem discussed in the next two chapters illustrates the power of experimentally manipulating a model system to understand tipping points and potentially intervene to forestall (or accelerate) regime shifts (Boettiger and Hastings 2012).

11.3 NEXT STEPS

Now that we have outlined some criteria to identify ecological tipping points and regime shifts, we will apply them to experimental work with the *Sarracenia* microecosystem. In chapter 12, we show that the pitcher fluid is either highly oxgenated or anoxic and that increases in a single environmental driver (detritus from excess prey) can lead to a rapid ecosystem collapse as measured by $[O_2]$ in the pitcher fluid. Because of the short generation time of the bacteria in the pitcher fluid, the *Sarracenia* microecosystem provides an ideal model for studying in

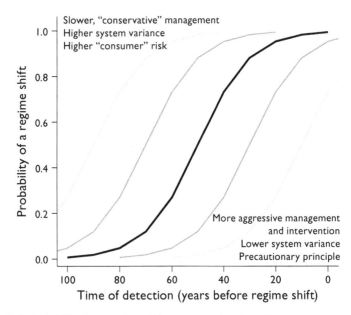

FIGURE 11.2. Probability that a regime shift occurs as a function of management response and the time of detection of an impending regime shift. As the rate of intervention declines or there is higher variance in the system (left side), earlier detection is required to avert a regime shift. With more aggressive or rapid intervention, or lower variance in the system (right side), less advance warning is needed to avert a regime shift. Modified from figure 9 of Contamin and Ellison (2009).

near-real time the long-term dynamics of nutrient enrichment and eutrophic collapse of aquatic ecosystems. In chapter 13, we use molecular tools to "downscale" to the proteomic level our understanding of ecosystem responses to enrichment. We observed a disconnection between higher-level measures of ecosystem function (i.e., $[O_2]$) and lower-level measures of proteomic diversity, leaving as an open question the meaning of the initial state of the *Sarracenia* microecosystem.

CHAPTER TWELVE

The Small World

Tipping Points and Regime Shifts
in the Sarracenia Microecosystem

Excitabat enim fluctus in simpulo.[1]
—Marcus Tullius Cicero (*ca.* 52 B.C.E. (2010 edition): 445)

Ecological assemblages, communities, and food webs have some amount of resistance to change caused by small, incremental changes in driver variables. For
"green" food webs that have a resource base of primary producers (algae or vascular plants), drivers often include nutrients— especially nitrogen and phosphorus—
that limit their growth. For "brown," detritus-based food webs like those found
in many small streams, larger rivers, pitcher plants, and bromeliads (chapter 10),
drivers may be allochthonous inputs of detritus. In both green and brown food
webs, these drivers can be thought of as ultimate or proximate variables that are
mechanistically responsible (sensu chapter 1, §1.4) for a substantial fraction of
the spatial or temporal variation in the state variables. In this chapter, we illustrate alternative states in the *Sarracenia* microecosystem and describe experiments
showing that excess nutrients (from prey capture) induce a tipping point and cause
a rapid shift from aerobic and anoxic conditions in the pitcher fluid. The shift
back from the anoxic to anaerobic state occurs much more slowly because microbial oxygen demand stays high as the detritus is decomposed. The experimental
tractability of the *Sarracenia* microecosystem, in concert with its dynamics that
are broadly representative of many other aquatic ecosystems, makes it an ideal
model system for studying ecological tipping points and regime shifts.

12.1 STATE CHANGES IN THE *SARRACENIA* MICROECOSYSTEM

The *Sarracenia* microecosystem is especially well suited to investigating
alternative states. For the brown inquiline food web found within *S. purpurea*
pitchers (chapter 9), the salient driver is arthropod prey that is captured by the
plant (Butler et al. 2008; Ellison and Gotelli 2009). The corresponding response
variable is the concentration of dissolved oxygen (O_2) in the pitcher fluid, which
reflects whether it is an aerobic or anaerobic state. When it comes to linking
empirical patterns with mathematical models of tipping points and alternative

states, the *Sarracenia* microecosystem has three clear advantages relative to other, larger aquatic ecosystems.

First, the *Sarracenia* microecosystem allows for true replication in a way that lakes, for example, do not. Each separate *S. purpurea* pitcher encloses and supports an independent, fully functional brown food web that includes a base of arthropod detritus, a diverse assemblage of microbes that break it down, and higher trophic levels of filter feeders and shredders (chapter 9; Miller et al. 2018). Leaves within plants and plants within bogs (figure 1.2; appendix A) are natural hierarchical levels of replication that cannot be found in other ecosystems.

Second, we can easily conduct controlled field and laboratory experiments with *S. purpurea* and its food web (chapter 9; Miller et al. 2018). Manipulative experiments are critical for establishing the link between candidate driver variables and their effects on temporal dynamics and avoiding false identification of tipping points and regime shifts (Boettiger and Hastings 2012).

Third, we can monitor changes in the drivers and responses of the *Sarracenia* microecosystem with frequent, nondestructive sampling (Sirota et al. 2013; Lau et al. 2018). Environmental changes that may take years or decades to manifest themselves in large ecosystems such as lakes occur within hours or days in *S. purpurea* (Sirota et al. 2013). A single eutrophication experiment that can be done in a week in a *Sarracenia* pitcher yields >10,000 temporally autocorrelated sampled values of O_2 concentration recorded at 60-second intervals. For a lake system studied for a decade, this would be comparable to taking measurements of dissolved oxygen every 8 hours, which, although possible, yields data much too slowly for making decisions (chapter 11, figure 11.2; Contamin and Ellison 2009). In short, studying tipping points and alternative states in the *Sarracenia* microecosystem yields rapid, cost-effective data and insights.

In other aquatic ecosystems, excess nutrient loading quickly swamps the capacity of higher trophic levels to control producer levels through grazing. Instead, the primary producers become self-limiting through shading, and biological oxygen demand rapidly increases as microbes break down the decaying plant material. We hypothesized that in the *Sarracenia* microecosystem there is the same limited capacity for higher trophic levels to control ecosystem processes in the face of excess inputs of prey.

12.1.1 Temporal Dynamics of Aerobic and Anaerobic Conditions in Sarracenia purpurea *Pitchers*

Figure 12.1 shows [O_2] in a single *S. purpurea* leaf measured at 60-second intervals over the course of a 10-day feeding experiment (Sirota et al. 2013). The resulting time series of 13,978 [O_2] data points (figure 12.1) is probably the most data-dense trajectory we have for collapse in any aquatic system; we can compare the patterns in this trajectory with the general predictions discussed in appendix F, §F.4.

At the start of the experiment, this leaf contained the full microbial component of *S. purpurea* pitcher fluid, but we removed all of the upper trophic levels (macrobial filter feeders, shredders, and top predators) to reduce some of the experimental complexity. Removing most of the food web may seem unrealistic, given the importance of top-down effects on trophic dynamics described repeatedly for the *S. purpurea* food web by many authors (chapter 9). However, the mineralization and translocation of N from prey into pitchers tissues occurs equally fast in the absence or the presence of the macroinvertebrates in the *S. purpurea* food web (Butler et al. 2008). That is, although the *S. purpurea* food web can be viewed as a processing-chain commensalism (Heard 1994), the microbes are doing the heavy lifting of detrital breakdown and nutrient mineralization, independent of the top-down control of microbial biomass in this system.

The trajectory of [O_2] in response to detrital loading had four phases (figure 12.1). During the preenrichment phase, [O_2] in the pitcher fluid fluctuated diurnally, reflecting the diurnal cycle of photosynthesis and respiration by the plant (Sirota et al. 2013; see also Bradshaw and Creelman 1984). As soon as we began adding dried, ground, and sterilized wasps (*Dolichovespula maculata*) as prey; the average [O_2] immediately began to fall even as it continued to cycle diurnally. Structural change analysis (Zeileis et al. 2003) identified a single breakpoint ≈ 77 hours after feeding began. We continued to enrich the system by feeding more ground wasps to the plant for an additional 5 days, during which time the average [O_2] was so low as to render the system anoxic, although the diurnal photosynthesis–respiration cycle persisted with a very small amplitude. After we stopped feeding the plant, we continued to monitor the system for 4 more days. Average [O_2] continued to drift downward, with no tendency to return to the pretreatment, oligotrophic state (figure 12.1; Sirota et al. 2013).

Appendix F lays out four a priori predictions for a system with alternative states: (1) multimodality in the frequency distribution of the responser variable; (2) rapid nonlinear change between alternative states; (3) critical slowing down preceeding a stage change; and (4) hysteresis following relaxation of the environmental driver. How did our experimental results compare with these predictions (appendix F, §F.4)?

Multimodality in State Frequencies   The data did fit a simple bimodal distribution in which the pitcher fluid was either well oxygenated or nearly anoxic, and the system spent relatively little time between those extremes (figure 12.2).

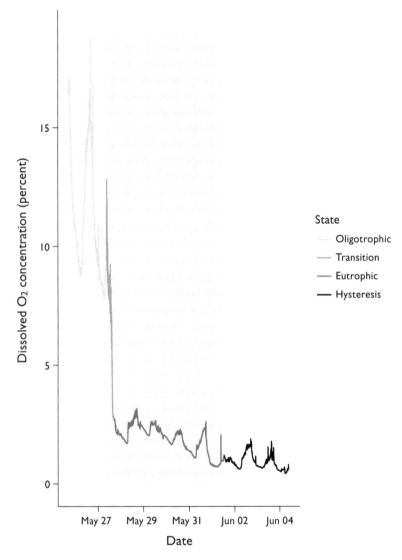

FIGURE 12.1. Dissolved oxygen concentrations [O_2] during eutrophication of the *Sarracenia* microecosystem. Over a 10-day period, [O_2] was monitored at 60-second intervals ($n = 13{,}978$ points); the gray-shaded box indicates the period of addition of excess prey in the form of ground wasps (*Dolichovespula maculata*). Increasing shading levels indicate the initial oligotrophic state, transition state, eutrophic state, and hysteresis after the termination of prey addition. Compare with predictions of the lower panel of appendix F, figure F.3.

The latter was confirmed by a simple sliding-window analysis of average values that identified a single breakpoint in this distribution. Although we described alternative states based on a single driver variable and a single response variable, the different states also could be visualized by plotting other measured variables. For example, a phase-plane plot for these data of photosynthetically active radiation versus [O_2] also illustrated two distinct clouds of parameter space separated by a thin link during the transition phase (figure 12.3).

Rapid Nonlinear Change between Alternative States   At first glance, this prediction appeared to be confirmed. However, note in figure 12.1 that the left-most part of the graph was the preenrichment phase. As soon as detritus was added (left side of gray-shaded box in figure 12.1), average [O_2] dropped sharply, although the diurnal periodicity was preserved during the transition. This pattern looks more like simple environmental tracking, with a steep decrease as soon as enrichment began, rather than a sudden nonlinear threshold that was reached after an initial period of enrichment.

Critical Slowing Down Preceding the State Change   The profile in figure 12.1 also showed little evidence for an increase in variance or autocorrelation ahead of the state change. Perhaps this pattern would have been more apparent in green food webs, where an initial input of limiting nutrients leads first to an increase in primary production before subsequent die-back of plants (algae) followed by increasing biological oxygen demand. In the brown food web of *S. purpurea*, [O_2] declined steeply and immediately as soon as we began adding additional prey.

Hysteresis Following Relaxation of the Environmental Driver   In the post-enrichment period of the experiment, the data certainly confirmed the predictions of hysteresis. Even though no additional detritus was added, [O_2] did not recover after 4 days (> 100 bacterial generations). In fact, average [O_2] continued to decline. However, without concurrent measurements of the driver itself, this pattern does not necessarily confirm a hysteretic response. For example, if the system had been overwhelmed by a high delivery rate, then the persistent low [O_2] simply may have indicated constant microbial activity as the bolus of detritus was gradually broken down and mineralized (Baiser et al. 2011).

12.1.2 An Alternative Approach

In a subsequent experiment, we replaced ground wasps with a soluble molecule—bovine serum albumin (BSA)—as our source of detritus. We loaded the pitchers

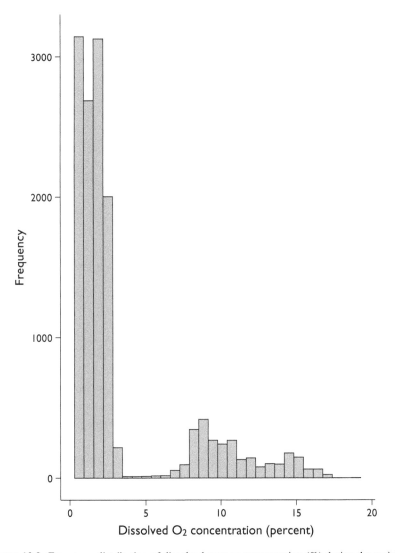

FIGURE 12.2. Frequency distribution of dissolved oxygen concentration (%) during the enrichment experiment illustrated in figure 12.1. Compare this result with the predictions shown in the lower panel of appendix F, figure F.2.

with BSA and simultaneously tracked the increase in BSA and the decrease in O_2. After the system collapsed to the eutrophic state, we stopped adding BSA and then monitored the recovery of O_2 as microbes transformed BSA and its concentration as an aqueous solute declined. We describe this experiment in much more detail in the next chapter.

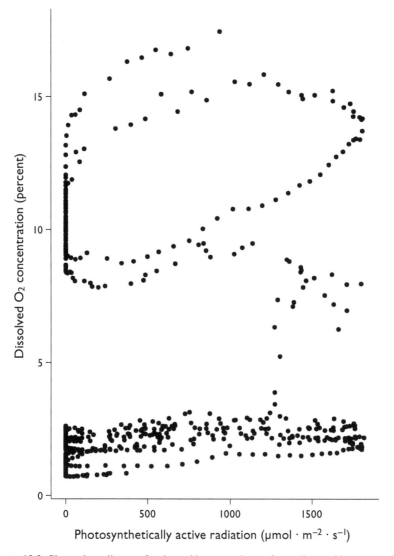

FIGURE 12.3. Phase-plane diagram for the sudden state change from oligotrophic to eutrophic conditions in figure 12.1. The *x*-axis is photosynthetically active radiation (PAR), which varies periodically through 24 hours, and the *y*-axis is dissolved oxygen concentration, which depends both on the time of day and on the nutrient regime. Each point is a consecutive observation in the 10-day time series. Note the thin band of points that separates the oxygen-rich oligotrophic state from the oxygen-poor eutrophic state.

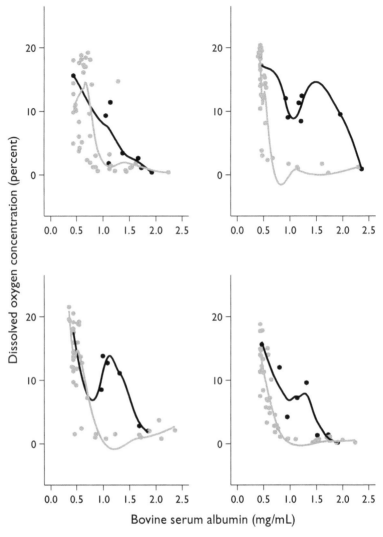

FIGURE 12.4. Replicated hysteresis experiments with the *Sarracenia* microecosystem. Each panel represents an independent replicate. The *x*-axis is the concentration of bovine serum albumin (BSA) that was added as a water-soluble nutrient source for microbes. The *y*-axis is the concentration of dissolved oxygen. Each point represents a different measurement during the enrichment phase (black points) and the recovery phase (gray points). The lines are separate LOESS splines fit to the enrichment (black) and recovery (gray) phases of the trajectory. Compare with appendix F, figure F.5.

For now, we simply highlight that the predicted pattern of classic hysteresis (figure F.5) was achieved in replicated *Sarracenia* microecosystems (figure 12.4). In these state-space diagrams, the driver variable (BSA) is on the *x*-axis and the response variable ([O_2]) is on the *y*-axis. We tracked both variables through time during the enrichment phase (black points) and the recovery phase (gray points). This graph confirmed the classic hysteresis pattern because [O_2] in the recovery phase did not achieve the same levels as during the enrichment phase, even when the BSA concentration was the same. As far as we are aware, this was the first demonstration of hysteresis in a controlled and replicated experiment in which only a single environmental driver (BSA) was incrementally increased until a tipping point was passed.

We return to these data in the next chapter to explore patterns in ways that have not been possible in the past but that are starting to be used by terrestrial ecologists who are willing to make space-for-time substitutions across environmental gradients. We will also show results for different nutrient delivery rates that reveal temporal dynamics in the *Sarracenia* microecosystem are far richer than we might have guessed from even simple descriptions of hysteresis. Finally, we will explore new work on metaproteomics, which may reveal candidate proteins and other biomarkers that could give us a better early warning indicators of state change than simply measuring oxygen.

12.2 SUMMARY

Replicated and properly controlled enrichment–recovery experiments as in figures 12.1 and 12.4 generate repeatable patterns and data-rich time series that provide strong tests for tipping points and alternative states. Even the simple profiles in figure 12.1 show the added complexity of diurnal fluctuations in oxygen concentration. These patterns and processes illustrate that the *Sarracenia* microecosystem provides ample opportunity to study the complex interplay between habitat structures, and ecological assemblages in slowly or rapidly changing environments.

CHAPTER THIRTEEN

Scaling Up

Using *omics to Identify Ecosystem States and Transitions

> *Existence is no more than the precarious attainment of relevance in an intensely mobile flux of past, present, and future.*
>
> —Susan Sontag (1969: 74–75)

In the previous chapter, we used *Sarracenia* as a model system for understanding tipping points, regime shifts, and the alternative aerobic and anaerobic states of the pitcher fluid and its microbial assemblage. In this chapter, we scale up from communities to ecosystems as we consider tipping points in the context of biodiversity and ecosystem function. At the same time, we "scale down" mechanistically to explore the use of proteins rather than species or operational taxonomic units (OTUs) to quantify ecosystem function. We find that metaproteomic profiles provide a more nuanced and satisfying measure of "ecosystem function" than more familiar measures of productivity, decomposition, etc. (appendix G, §G.1). Intriguingly, we find that one higher-level measure of ecosystem function in the *Sarracenia* microecosystem—oxygen content of the pitcher fluid—recovers from an induced anaerobic state, but its proteomic profiles do not. This result leads us to reflect on whether this system has really returned to "the same" initial state (appendix F, figure F.1) and suggests future directions for research in tipping points and regime shifts.

13.1 PROTEIN SURVEYS OF THE *SARRACENIA* MICROECOSYSTEM

Proteins are basic constituents of cells and tissues, so a proteomic survey of an entire ecosystem would include a profile of the proteins in the component species (appendix G, §G.2.1). In a proteomic survey of three of the obligate pitcher-plant inquilines (*Wyeomyia smithii*, *Metriocnemus knabi*, and *Fletcherimyia fletcheri*), we identified more than 50 proteins from each species, 10 of which were predominantly or uniquely found in a single species (Gotelli et al. 2011). Using an assembled database of 100 metazoan myosin heavy-chain orthologs, we identified 19 peptides unique to one of the 3 species. In a future version of the classic Star

Trek "tricorder"[1] (or a soon-to-be app for our smartphones or watches), these unique molecular signatures could be used to recognize the presence of proteins, quantify their abundance, or diagnose a person's health from real-time proteomic surveys of biological samples (or the wearer's sweat; Sarwar et al. 2019).

In many communities, from the human gut (Human Microbiome Project Consortium 2012) to acid mine drainages (Ram et al. 2005; Méndez-García et al. 2015) and terrestrial soils (Wagg et al. 2014), it is the activity of microbes that translates into much of what we ultimately characterize as ecosystem function. In nutrient-enriched aquatic ecosystems, it is the breakdown by microbes of accumulated dead plant material that ultimately causes eutrophication—the depletion of oxygen and subsequent change in biodiversity to favor species that can tolerate low oxygen levels (Carpenter et al. 1998).

13.2 PROTEOMICS OF *SARRACENIA* FED SUPPLEMENTAL PREY

To explore ecosystem function in the *Sarracenia* microecosystem, we have focused on its microbiome and its array of sampled and sequenced proteins. We began our work by sampling microbes from intact *S. purpurea* pitchers and constructing a microbial metagenome that we used as a scaffolding for protein identification. Next, we set up a field experiment to assay protein diversity in enriched and control pitchers (Northrop et al. 2017).

Methods   Over a 5-day period in June 2011, we selected 20 pitcher plants (10 experimental, 10 control) in the bog at Tom Swamp in Petersham, Massachusetts (Swan and Gill 1970). The newly opened pitchers of the control plants were fed daily for two weeks with 1 mg/mL of oven-dried, sterilized, finely ground wasps (*Dolichovespula maculata*). Newly formed pitchers are sterile until they open (Hepburn and St. John 1927; Peterson et al. 2008), so all microbes and macrobes have to colonize them afresh (chapter 9, §9.4). We did not exclude invertebrates from colonizing the experimental plants, but the fluid we analyzed for proteomes was filtered to isolate a microbial pellet from each plant. For five of the control plants and six of the fed plants, there was sufficient material for gel electrophoresis and tandem mass spectrophotometry. Preliminary analyses did not reveal any microbial proteins in the ground wasps, and their gel electrophoresis bands were very different from the bands of electrophoresed microbial pellets from fed pitchers (Northrop et al. 2017).

At the end of the experiment, we used shotgun metaproteomics with a custom metagenomic database to identify proteins that were mapped back to 86 named bacterial species. Because of the impossibility of counting even a small fraction of all the possible proteins in an ecosystem, our strategy instead was to classify and estimate the abundances of the most common proteins in each treatment group, regardless of their absolute abundances. For the *S. purpurea* microbiome, this turned out to be 220 proteins in each treatment that we could sequence and identify reliably. Each of the these 220 identified proteins in each treatment, ordered by the number of total peptides associated with the protein hit, was matched to a protein name. We eliminated duplicate names and used other statistical safeguards against false positives and misidentifications. We interpreted the total number of peptides associated with each protein as a crude index of the relative abundance of each protein in the sample (Northrop et al. 2017).

Results   The metagenome was dominated by bacteria (99.1%), with slightly higher diversity in the control pitchers (69 named species in 12 bacterial classes) than in the fed pitchers (53 named species in 11 bacterial classes). Betaproteobacteria was the most common class in both treatments (see also Koopman et al. 2010), but with higher representation in enriched (84.4%) versus control (50.3%) pitchers (Northrop et al. 2017).

The protein profiles of the two treatments were quite distinct. There were only 65 proteins shared between the two treatments, with 155 unique proteins in the controls and 155 other unique proteins in the fed treatment. If the null hypothesis of no treatment effect were true, these numbers would have been reversed, with approximately 155 shared proteins expected by chance. We ordered the top 20 most common proteins in each treatment by the number of peptide hits and constructed the equivalent of rank–abundance curves for them. As in a species rank–abundance curve, there were a few common proteins and many rare ones in each of the samples. Even restricting ourselves to the top 20 in each treatment, the drop-off in relative abundance from the most to the least common protein was about 18-fold, and this number would have increased by many orders of magnitude if we had continued sampling from the long right-hand tail of increasingly rare proteins (figure 13.1).

Given that only 65 proteins were shared in the two lists of the top 220, we fully expected that none or few of the top proteins would be the same in the two lists. We also expected the protein composition of the most common elements to be different in the two treatments, because this is what we always observe when we compare the composition of macrobial species assemblages. For example, in freshwater habitats, eutrophic and oligotrophic assemblages reliably contain different sets of the most common indicator taxa (Vaughan and Gotelli 2019). The characteristic changes in the common species and their functional traits are so reliable and familiar that many water-quality indices and restoration objectives are set by the presence of these indicator taxa (Poff et al. 2006). Similar differences in taxonomic diversity and metabolic pathways of microbes are representative of proteomic and

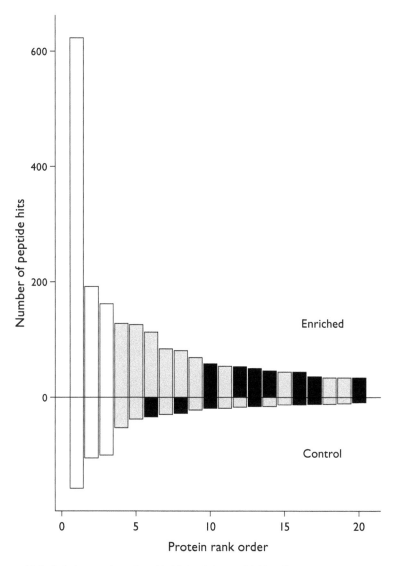

FIGURE 13.1. Relative numbers (peptide hits) of the top 20 identified proteins in control and enriched *S. purpurea* pitchers. Unfilled bars: proteins with the same rank in control and enriched treatments; gray bars: proteins with differing ranks in control and enriched treatments; black bars: unique proteins that were identified in only one of the two treatments.

microbial responses to nutrient enrichment in freshwater ecosystems (Haller et al. 2011; Kearns et al. 2016).

But this pattern of characteristic dominant groups did not hold for the proteins we recovered from *S. purpurea* pitcher fluid. Instead, we discovered that the three most abundant proteins in both control and enriched pitchers—"EF-Tu" (elongation factor thermo unstable), "FOF1 ATP synthase subunit beta," and a "molecular chaperone"—were identical. The next two most abundant proteins ("branched chain amino-acid ABC transporter substrate-binding protein," and a "porin") differed by only one rank between the control and enriched pitchers. Collectively, these identical proteins accounted for about 60% of the peptides in each of the treatments (figure 13.1; Northrop et al. 2017).

It was the more intermediate proteins that differed between the two treatments. For the next 15 proteins in the list, there were 7 unique proteins in each treatment. The remaining 8 proteins were found in both treatments, but usually with very different ranks in abundance. Why should the most common proteins in both lists have been nearly identical when the two sets differed sharply in overall composition? The most likely explanation was that these 5 proteins are part of evolutionarily conserved pathways for critical cellular functions. Although protein profiles do ultimately map back in some ways to species identity, there is not a simple congruence with other measures of species diversity.

13.3 THE CYBERNETICS AND INFORMATION CONTENT OF THE *S. PURPUREA* PROTEOME

With the protein data in hand, we returned to the cybernetics of ecosystems and ask whether there was a difference in the information content (measured as Hurlbert's Probability of Interspecific Encounter [PIE]) of the proteomes of the control and enriched pitchers.[2] Despite the distinct protein composition of the intermediate ranks, PIE was almost identical in the two treatments (control PIE = 0.94, enriched PIE = 0.92), so there was no evidence that the relative evenness of these distributions differed between the control and fed pitchers. However, we obtained a different result when these same proteins were mapped back to different taxonomic groupings. Although there were comparable numbers of bacterial classes detected in the two treatments (control = 12, enriched = 11), the taxonomic PIE for the control plants (0.71) was significantly higher than in the enriched plants (0.31), indicating a more even relative abundance distribution in the controls (figure 13.2; Northrop et al. 2017).

As with the protein identification, there were important shifts in composition: aerobic bacteria (Betaproteobacteria) were more common in control pitchers

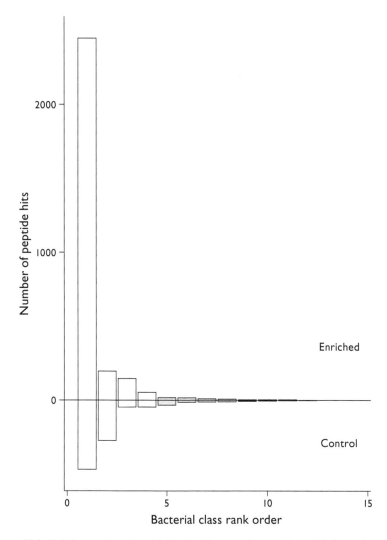

FIGURE 13.2. Relative numbers (peptide hits) of the most frequent bacterial classes in control and enriched *S. purpurea* pitcher fluid. Taxonomic identifications were constructed by mapping identified proteins (figure 13.1) back to a metagenomic library previously constructed for *S. purpurea*. Symbols and shading as in figure 13.1.

and facultatively anaerobic bacteria (Alphaproteobacteria) were more common in enriched pitchers. Finally, we detected some differences in KEGG molecular pathways of nucleotide and energy metabolism. In sum, distinct differences at the protein and peptide levels corresponded to differences in molecular pathways and taxonomic differences at the genus, family, and class levels (Northrop et al. 2017).

13.4 EARLY WARNING INDICATORS, HYSTERESIS, AND THE
TWISTED PATH OF FUNDED RESEARCH

Scientific research rarely happens as we describe it in our research proposals or even in the subsequent papers we write. Based on our discovery that proteomes were distinctly different for control and enriched *Sarracenia* assemblages (Northrop et al. 2017), we received funding from the US National Science Foundation funding to search for proteins that might serve as early warning indicators that a system was on its way to collapse (chapter 11, figure 11.1). The hope was that key proteins that reflect changes in microbial activity might give us more lead time for intervening in aquatic ecosystems nearing a tipping point.

In a nutshell, we proposed to use supplemental prey to enrich *Sarracenia* microecosystems, track key bioindicator molecules with real-time proteomic screens, stop feeding immediately on detecting reliable indicator molecules, and test whether our intervention (the "stopping rule") would stop the impending collapse. Unfortunately, it didn't work out that way. We thought that by varying the rate of prey addition, we could control the rate of collapse in $[O_2]$. To our dismay, we found that the system either collapsed within 8 hours of the initial prey addition or did not collapse at all.

This observation was important because it showed that the *Sarracenia* microecosystem switched rapidly between high and low $[O_2]$ but spent little time at intermediate concentrations. This kind of bimodality in the distribution of a continuous response variable is one of the criteria used to recognize a system with alternative states (appendix F, figure F.2). But it meant that we had insufficient human time to use real-time proteomic assays to cue and initiate an intervention.

What to do next? We had already shown from press experiments that the temporal trajectory of $[O_2]$ changed with increased prey addition (figure 13.3; Sirota et al. 2013). With no or little prey addition, $[O_2]$ exhibited a strong diurnal cycle with peaks during diurnal photosynthesis and troughs during nocturnal respiration. With increasing prey, these cycles broke down and the system exhibited more complex and unpredictable dynamics. At the two highest levels of prey addition, $[O_2]$ collapsed quickly. After the collapse, a diurnal $[O_2]$ cycle was restored, but at much lower levels and with a greatly reduced amplitude. In a series of pilot pulse experiments in which we briefly enriched the system, we noticed that the dynamics after the collapse also were affected by the size of the enrichment pulse. Collectively, these results suggested that it might be profitable to examine the "back end" of eutrophic collapse and investigate hysteretic responses of *S. purpurea* pitchers as they recovered from anaerobic conditions.

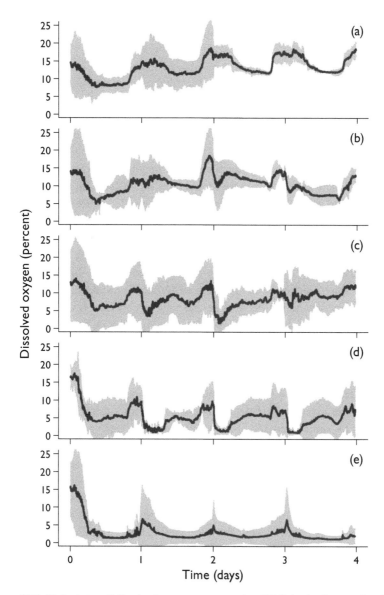

Figure 13.3. Trajectories of dissolved oxygen concentration ($[O_2]$) in the *Sarracenia* micro-ecosystem exposed to different enrichment regimes. Each panel depicts the average (black line) and 95% confidence interval for six replicate ecosystems. From top to bottom, the enrichment rates were 0 (control), 0.125, 0.250, 0.500, and 1.000 mg $\cdot$ mL^{-1} $\cdot$ d^{-1} of dried, ground wasps (*Dolichovespula maculata*). In the control and low-enrichment treatments (**a, b**), there was a daily periodicity to $[O_2]$ reflecting the diel cycle of plant photosynthesis and respiration. At the highest concentration (**e**), there was a collapse of $[O_2]$ followed by a diel periodicity but with a greatly reduced amplitude. Redrawn from Sirota et al. (2013).

13.4.1 Hysteresis, Environmental Tracking, and Anti-hysteresis in the Sarracenia Microecosystem

A clear demonstration of hysteresis in ecological systems has not been easy, in part because of the long duration of ecological change in most ecosystems. One of the best examples comes from experimental grassland plots in which nitrogen addition was halted, but vegetation composition and structure still had not converged 20 years later (by 2011) on that of the unfertilized control plots (Isbell et al. 2013). But in that study, there also was ongoing climatic warming during the two-decade recovery period that could have interacted with the experimental treatments and altered the return trajectory of the plots after fertilization was stopped.

The *Sarracenia* microecosystem seemed ideal for an experimental study of hysteresis. We could easily push the system from an aerobic to an anaerobic state by adding prey or inorganic nutrients and monitor subsequent changes in $[O_2]$ with microelectrodes. A signature of hysteresis would be a difference in the $[O_2]$ trajectory during the decline and recovery phases. But a new problem arose as we started to design these experiments. Although we could easily monitor $[O_2]$, we could not monitor the amount of remaining detritus—the driver variable (Lau et al. 2018). Measurements of undigested ground wasps were not feasible because the material was difficult to extract and weigh accurately, and a proteomic assay would require destructive sampling to get enough material. For more than 6 months, we experimented with alternative delivery schemes, including using dried ants (a natural resource packet) and even "tea bags" of ground wasps that were steeped in the *S. purpurea* "teapot." But all of these measurements were either too disruptive to the system or simply too noisy to give accurate estimates of undigested detritus.

Our *omics collaborator Bryan Ballif solved the problem. He suggested that instead of using natural sources of detritus, we switch to bovine serum album (BSA), a nitrogen-rich soluble molecule that microbes readily consume and transform. By adding additional DNA salts, we were able to create an aqueous molecular cocktail that had nutrient ratios similar to those of natural prey. We confirmed with pilot experiments that the collapse and recovery of $[O_2]$ in *S. purpurea* pitchers laced with BSA was similar to the $[O_2]$ profile generated by feeding pitchers with whole-insect prey or ground wasps. We also established that BSA was a stable molecule that did not break down or decompose except in the presence of microbes. BSA concentrations in water can be accurately measured from microliter samples using the Bradford assay (Zor and Seliger 1996), so we could frequently monitor the remaining undigested levels of BSA in our system while simultaneously monitoring $[O_2]$ with a microelectrode.

Methods   We established replicate pitchers assigned to three treatments: low, medium, and high rates of BSA addition (0.5, 2.0, or 5.0 mg BSA · mL^{-1} (Northrop et al. 2021). At the start of the experiment, each pitcher was seeded with an aliquot from a common stock of *S. purpurea* pitcher fluid that had been passed through a 30-μL frit bed of a chromatography column filtered to remove all macrobes and detritus. We added BSA, monitored [O_2] and BSA until the system collapsed, then stopped adding BSA and continued to monitor until [O_2] returned to approximately their starting values (about 20% for a typically well-oxygenated *Sarracenia* leaf).

Results   The trajectories of collapse and recovery were repeatable for different pitchers that received the same treatment (chapter 12, figure 12.4). But the trajectories differed in a predictable way based on the BSA treatments (figure 13.4; Northrop et al. 2021). In the low-BSA treatment, the trajectory followed a classic pattern of hysteresis. After collapse and the termination of BSA additions, [O_2] remained low even though BSA levels had declined from microbial activity. [O_2] did not begin to rise again until BSA had declined to fairly low levels. The pattern is reminiscent of many large aquatic systems in which N and P additions have been reduced or eliminated, but the system fails to recover rapidly to a "clean" (oligotrophic) regime with high levels of oxygen and biodiversity.

In the intermediate BSA treatment, there was no evidence of a hysteretic response. After we stopped adding BSA, the [O_2] profile retraced its steps in parallel with the decline in BSA. This trajectory characteristic of systems in which the response variable quickly tracks the environmental driver with little or no time lag in the response to changing levels of the driver variable (appendix F, figure F.1).

In the high-BSA treatment, we discovered a pattern of "reverse" or "counterclockwise" hysteresis: [O_2] during the recovery phase was actually higher than it was during the collapsed phase (Northrop et al. 2021). This pattern of reverse hysteresis has been recognized previously in clinical pharmacology trials (Louizos et al. 2014), sediment discharge studies (Williams 1989; Evans and Davies 1998), heart rates in lizards during heating and cooling treatments (Grigg and Seebacher 1999), and in a Chinese floodplain (Zhang and Werner 2015). But the anti-hysteresis we observed in *S. purpurea* pitchers is the first example of this kind of dynamics in a controlled and replicated ecosystem experiment in which treatments differed only in the delivery rate of the environmental driver (Northrop et al. 2021).

Insights   Our hysteresis experiments provided several new insights. First, even in a replicated, controlled experiment in which we varied only the delivery rate of a microbial substrate, the resulting dynamics were far richer than had been described previously. The diversity of dynamic responses in this simple experiment may

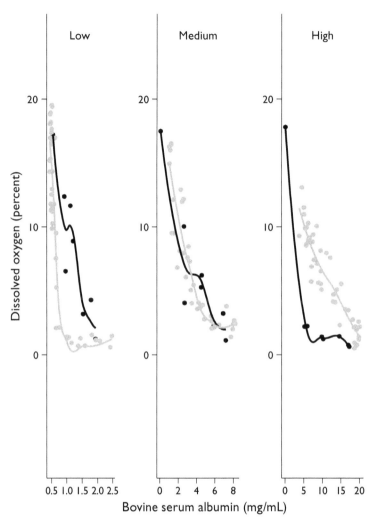

FIGURE 13.4. Hysteresis dynamics of the *Sarracenia* microecosystem. Each panel represents the average values for replicated pitchers in different experimental treatment, with Low (0.5 mg/mL), Medium (2.5 mg/mL) and High (5.0 mg/mL) rates of daily enrichment with bovine serum albumin (BSA). For the first 4 days of the experiment, pitchers were enriched once a day, with $[O_2]$ measurements taken with a microelectrode immediately after enrichment. Standing concentrations of BSA and $[O_2]$ were measured twice daily during days 0–20 of the experiment and once daily for days 20–28. Pitchers were monitored until $[O_2]$ levels returned approximately to starting conditions before nutrient enrichment. The Low treatment exhibited a classic pattern of hysteresis, the Medium treatment exhibited a pattern of environmental tracking, and the High treatment exhibited a pattern of reverse or counterclockwise hysteresis. Redrawn from Northrop et al. (2021); points, shades, and lines as in Fig. 12.4.

explain why some restoration projects at larger scales have failed. The outcome of a restoration project may depend not only on the current levels of an environmental driver, but rather on the past history and levels of the driver during the preintervention phase (Jones and Schmitz 2009). Specifically, our results suggest that systems that have been exposed to persistent low levels of an environmental driver may be especially difficult to restore. For this reason, long-term records of environmental monitoring may be important for predicting the success of restoration projects.

Second, although it is not apparent from the graphs in figure 13.4, both the $[O_2]$ and the time scales are very different in the recovery phase of the trajectories. At low and intermediate BSA levels, the return time to a fully oxygenated state occurred in ≈ 15 days. But in the high-BSA treatment, full recovery did not occur until 35 days, and a few pitchers still had not fully recovered by the time that we terminated the experiment.

Finally, despite their differing trajectories, the pitchers in the different treatments eventually returned to initial $[O_2]$. But had they actually returned to the same initial state? With a proteomic assay, we confirmed that the pitchers in all three treatment groups did have the same microbial profile at the start of the experiment. But by the end of the experiment, the taxonomic profiles of microbes identified from the proteomic assay differed significantly among the three treatments (figure 13.5). One hypothesis to explain this result is that there was some taxonomic redundancy in the microbial assemblages and the same functions were being provided by different taxa in the different assemblages. Alternatively, the imprint of the past was lost and ecosystem function was no longer the same, even though $[O_2]$ levels were restored. A definitive experiment here would be to run the treatments through a second cycle of enrichment and recovery, with further proteomic assays at different points in time. We will leave that experiment—and others described in the next chapter—to future generations of ecologists.

13.5 SYNTHESIS

Much current ecological research is focused on defining and measuring ecosystem functions, but most of their definitions reflect categories of ecosystem services that are beneficial for human health and well-being. Proteins are biologically important molecules for all physiological functions of organisms, and the array of proteins generated by an entire ecosystem (the metaproteome) may be a useful way to characterize ecosystem function. Because microbes are responsible for the decomposition of detritus, expressed proteins could serve as potential biological indicators of impending regime shifts in enriched aquatic systems. In enriched and control *S. purpurea* pitchers, the microbial metaproteome differed significantly in

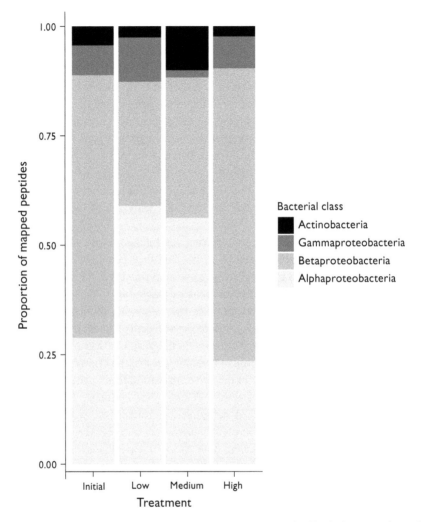

Figure 13.5. Taxonomic profiles of common *Sarracenia* bacterial families in three experimental treatments of Low, Medium, and High rates of BSA enrichment. Although all three treatments started with the same microbial composition, and [O_2] had returned to control levels, the three treatments were taxonomically different from one another.

the identity of proteins that could be mapped back to different microbial taxa and cellular pathways. An additional enrichment experiment revealed that taxonomic diversity derived from protein assays differed among treatments that had recovered to an aerobic state. This suggests that *omic biodiversity and ecosystem function may be decoupled and should encourage further studies to determine the best measures or levels of biodiversity to link to important ecosystem services.

CHAPTER FOURTEEN

Conclusion

Whither Sarracenia*?*

> *We perfected and practice [f]oretelling ...[t]o exhibit the*
> *perfect uselessness of knowing the answer to the wrong*
> *question.*
> —Ursula K. Le Guin (1969: 57)

> *Forty-two.*
> —Deep Thought, in Douglas Adams (1980: 180)

In chapter 1, we described three uses of scale in ecology: (1) experiments and observations conducted at different spatial and temporal grains and extents (Levin 1992); (2) scale-free analyses of dimensionless parameters that describe constant relationships between variables (West et al. 1997); and (3) model systems such as *Sarracenia* for which results can be scaled up to other systems that operate on longer time scales and are harder to manipulate in replicated experiments Ellison et al. (2003). In this closing chapter, we take stock of the ground we have covered and make suggestions for future research.

Most of our work on scaling fits within Levin's (1992) framework. We have scaled hypotheses, methods, and results up or down by changing the grain and extent of spatial and temporal observations, measurements, and experiments. Some of our analyses of stoichiometry and food-web structure have been, if not dimensionless, at least scale free (Brown et al. 2004). For example, in chapter 5 we asked whether nutrient stoichiomettry of *Sarracenia* leaf tissue was consistent with the global spectra of plant traits (Díaz et al. 2016). And in chapter 11, we asked whether inquiline food webs of *Sarracenia* conformed to Ulanowicz's "window of vitality" (Ulanowicz et al. 2014).

More generally, we have tried throughout this book to make the case that the *Sarracenia* microecosystem—the plant *S. purpurea* and its inquiline food web—is a true model system for ecological research. Like the more familiar model organisms used in laboratories around the world, *S. purpurea* and its inquilines are easy to propagate and grow in the lab and in the field. The inquilines, especially the protozoa and bacteria, have short generation times, and their growth is responsive to resources, abiotic conditions, and interactions with each other and with the plant. As genetic data and phylogenies of *Sarracenia* and its inquilines become more readily available (Ellison et al. 2012; Merz et al. 2013), it should be easy to maintain stock cultures of genetic lines of them that exhibit or lack key traits, just as we currently do for fruit flies, mice, roundworms, and cress.

14.1 RESOURCES, NUTRIENTS, AND STOICHIOMETRY

Some of our earliest work with *Sarracenia* focused on prey and how the plant responds to it. We documented substantial morphological plasticity along the pitcher-to-leaf continuum as a function of nitrogen addition (Ellison and Gotelli 2002). This result mirrored Mandossian's 1966 results for the morphology of *S. purpurea* along a sun-shade gradient and were replicated by Bott et al. (2008) in a Wisconsin bog. However, Lindsey Pett, a PhD student at the University of Vermont in Nick Gotelli's lab, repeated the nitrogen and phosphorus addition experiment at Molly Bog in the summer of 2018. Although some of her treatment levels were within the bounds used by Ellison and Gotelli (2002), she did not detect significant differences in keel morphology of plants fed additional N. We don't think that this result was because of differences in nutrient addition treatments or experimental execution. Rather, Molly Bog has been subject to 20 additional years of N loading since our first experiment; we hypothesize that habituation of old plants to chronic N loading and strong directional selection on young plants may be reducing plastic responses to nutrient loading.

Treating prey as a packet of resources for the plant, we have been able to ask how *S. purpurea* and other carnivorous plants fit into the global spectra of leaf traits and nutrient ratios measured in other terrestrial plants (Díaz et al. 2016). Although some carnivorous plant species differ greatly in their tissue nutrients from most terrestrial plants, the genus *Sarracenia* does not (Ellison 2006). Similarly, when *Sarracenia* was supplemented with additional prey, the shifts in nutrient allocation stayed within the ranges observed for other plant species (Farnsworth and Ellison 2008). However, when *Sarracenia* was supplemented with inorganic nutrients, the stoichiometric "rules" were broken, at least in the short term (chapter 4).

Two competing hypotheses in litt. for understanding stoichiometric ratios are homeostasis and the "you are what eat" hypothesis (chapter 1, appendix B). Scaling these hypotheses from individuals to populations suggests parallels with population regulation and density dependence (Sibly and Hone 2002). Homeostatic control of nutrients should result in relatively constant nutrient ratios with limited variation around an equilibrium point. This constancy is analogous to the density-dependent birth and death rates that limit fluctuations in population sizes of plants and animals. Gotelli et al. (2017) scaled up population-level regulation one step further. They found evidence for regulated trajectories of community-level species richness and total abundance in a variety of assemblages that had been monitored annually for more than 10 years. In contrast, the more flexible stoichiometric ratios that reflect "you are what you eat" are equivalent to expectations of a null model of an unconstrained random walk, either for population growth or for Gleasonian succession (appendix E).

Can these concepts of homeostasis and random walks be scaled up to entire ecosystems (Patten and Odum 1981)? In collaboration with Angélica González (Rutgers University), we are starting to examine these competing hypothesis in all the different nutrient "pools" of the *Sarracenia* and bromeliad microecosystems: leaf tissue, pitcher water, prey and detritus, microbes, and macrobes. Using press and pulse experiments, we are experimentally altering nutrient inputs to pitcher water and bromeliad tank water and then measuring stoichiometric ratios of these nutrient pools in the same plant. We are asking (1) How much do nutrient ratios fluctuate in these different pools? (2) Are some pools stable and homeostatic whereas others are more labile and variable?

Answering these questions will require careful comparisons with an appropriate null hypothesis. As a first pass, we are sampling additional technical replicates (subsamples of pooled replicates; see appendix G, §G.2.3) to estimate the fraction of variability in each nutrient pool that simply represents measurement error. We are combining these experimental studies with measurements of nutrient profiles in *Sarracenia* (chapter 3) and in arboreal bromeliads (§3.2; González et al. 2011) along a geographical gradient from Canada to Brazil. This gradient encompasses great variation in rates of atmospheric deposition and potentially in the relative importance of P versus N limitation (Reich and Oleksyn 2004). Detailed knowledge of the food webs of *Sarracenia*, bromeliads, and other cavity-dwelling assemblages (chapter 10) across geographical gradients also will improve our ability to scale the dynamics of resource use across levels of biological organization.

14.2 DEMOGRAPHY AND SPECIES DISTRIBUTIONS

Demography is at the root of population biology (Egerton 2012), and forms the basis for simple projections of population growth. Our demographic work on *S. purpurea* started with classical methods of tracking individuals within cohorts, measuring rates of dispersal, establishment, and density dependence, and incorporating demographic vital rates into matrix population projection models (chapter 6). We wanted to use these basic demographic models to predict the risk of extinction in response to environmental drivers such as atmospheric nutrient deposition. However, most current methods for forecasting fates of populations in response to environmental change ignore demography. Instead, they use geo-referenced specimen records combined with data layers of down-scaled environmental variables estimated for current or future conditions to forecast future habitat suitability (appendix D; Guisan et al. 2017).

In chapter 6, we proposed an alternative methodology based on demography: (1) experimentally manipulate potential environmental driver variables (e.g., warming, nutrient deposition) in small-scale field experiments; (2) measure life-table responses of demographic vital rates to differing levels of the driver; and (3) construct stochastic or deterministic models of population growth to estimate λ, the projected rate of population growth as a function of the driver variable. In chapter 7, we scaled this approach up to the North American continent. We applied the projection model we had built for Molly and Hawley Bogs (two pixels in the map of North America) to the native range of *S. purpurea* in the United States by combining the life-table response functions with continental measures and future projections of atmospheric N and P deposition.

This framework could be expanded and improved upon in at least four ways. First, individual and population-level responses to the same environmental driver may depend on the position of the population within its geographical range (Hampe and Petit 2005). Thererfore, small-scale experiments should be conducted at the center and at the margins of a species geographical range. Second, we should gather information on the frequencies of different genotypes within and between populations, which may improve the forecasts of demographic and species distribution models (Clark et al. 2011; Gotelli and Stanton-Geddes 2015). Common-garden experiments that bring together a range of genotypes collected throughout the geographical range of a species would also be informative (Whitham et al. 2006). Third, we should incorporate density dependence into demographic models, which may generate more realistic estimates of long-term population persistence (Ehrlén and Morris 2015). Finally, we should integrate direct effects of environmental drivers on individual physiology and population growth with the indirect cascading effects of changes in the abundance of other species in the assemblage (Diamond et al. 2016). Attention to these issues may yield more insights than further statistical refinements of species distribution models built from only specimen occurrence records and a few climatic variables.

14.3 FOOD WEBS AND OTHER NETWORKS

The starting point of our research on the *Sarracenia* microecosystem was the inquiline food web and its mutualistic interactions with the plant. These interactions also reflect the property that the *S. purpurea* food web assembles and develops in a living habitat that itself is changing on a comparable time scale. Although this is not a universal property of food webs, the network properties of the *S. purpurea* food web are similar to those of a wide range of more typical trophic and biogeochemical networks from more "static" habitats (chapter 10). Already in 1997, we were designing experiments to add a temporal dimension to food-web assembly (chapter 9; Ellison et al. 2003) with an eye toward linking the three major strands of community ecology: co-occurrence and assembly rules, food webs, and succession (chapter 8 and appendix E; see also Blanchet et al. 2020). As with our analyses of *Sarracenia* nutrient stoichiometry, our work on its dynamic food webs suggested ways to scale from individuals to ecosystems and to identify structural network properties that are independent of the level of biological organization.

Much research in community ecology has emphasized organismal traits, phylogenetic constraints, and temporal dynamics. Over the past two decades, traits and phylogeny have been steadily incorporated into analyses of classically defined "communities"—taxonomic or functional guilds operating at a single trophic level. Much more recently, interest has emerged in temporal analysis of "multilevel" food webs and other networks. Simultaneously, sequencing of entire assemblages—metagenomics (a.k.a. community genomics)—has been revealing the true diversity of microbial assemblages and providing much finer taxonomic resolution of the microbial compartments of food webs. Despite these advances, community ecologists continue to make a somewhat artificial distinction between the static, abiotic "habitat" and the dynamic biotic community it contains. Now is a good time to move beyond that dichotomy.

The *Sarracenia* microecosystem is an excellent system for research in all of these advancing avenues. Several research groups are currently focused on the metagenomics of the microbial inquilines in several *Sarracenia* species (Koopman et al. 2010; Satler et al. 2016; Grothjan and Young 2019). Much past research on the inquiline food web has focused on top-down control by the *Sarracenia* macroinvertebrates but within a few years, we will be able to tease out the role of bottom-up processes controlled by microbes. Specifically, it should soon be possible to construct custom microbial food webs in *S. purpurea* either by selective addition of specific taxa (OTUs) or selective deletion of them using targeted antibiotics or fungicides. Although the microbial component of the *S. purpurea* food web is the primary engine of nutrient mineralization and immobilization (Butler et al. 2008), we still do not know whether or how the plant itself competes with microbes for macro- and micronutrients.

The phylogeny of *Sarracenia* has been resolved reasonably well (Ellison et al. 2012), and Jonathan Millett is leading an effort to build a database of traits for *Sarracenia* and other carnivorous plants; an initial version of this database, including

127 species and 17 traits, has been posted to TRY. Parallel efforts for the macrobial and microbial inquilines would be of great use. In particular, we need more data for the apparently co-speciating clade of *Sarraceniopus* mites (Naczi 2018), the group of sarcophagid flies that co-occur with multiple *Sarracenia* species (Dahlem and Naczi 2006), and the rotifers and microbes that occur in *Sarracenia* and its surrounding habitat (Błędzki and Ellison 2003). Open questions include (1) How does a dynamic (living and changing) habitat act as a "filter" for trait-based community assembly? (2) Can we add habitat dynamics as another dimension into models of multilayer networks? (3) Does the inclusion of intraspecific variability alter food-web dynamics in the *Sarracenia* microecosystem, as it appears to in many others (Poisot et al. 2015)? To link some of these questions to evolutionary dynamics and longer time scales, a map of the quantitative trait loci of prey-capture traits (Malmberg et al. 2018) and the phylogeny of *Wyeomyia smithii* associated with its postglacial range expansion (Merz et al. 2013) will provide key data.

14.4 TIPPING POINTS, REGIME SHIFTS, AND ALTERNATIVE STATES

Although tipping points and regime shifts are still on the au courant bandwagon, theoretical work has far outstripped the data available to test them (Petraitis 2013). Indeed, our experimental work with tipping points and alternative states was motivated by modeling studies of lake and ocean ecosystems, which suggested that current methods for identifying early-warning indicators are woefully inadequate for managing ecological change (Biggs et al. 2009; Contamin and Ellison 2009). In the *Sarracenia* microecosystem, we could replicate entire ecosystems, manipulate a single environmental driver, and continuously monitor the system to rigorously identify alternative states. Most importantly, we could generate the densely sampled, autocorrelated time-series data required by the models, but that are rarely collected by ecologists (Bestelmeyer et al. 2011). We used the *Sarracenia* microecosystem to scale down from ecologists' conventional, macroscopic view of system-level tipping points and state changes to begin to develop a mechanistic understanding of these processes at *omic levels.

Defining potential alternative states and determining whether changes between them are abrupt or smooth are the first two challenges in the study of tipping points and regime change. The *Sarracenia* microecosystem was of great value here because, like many other freshwater and marine ecosystems, it has only two states that are relatively straightforward to identify: a well-oxygenated, oligotrophic state and an anoxic, eutrophic state.[1] Despite this apparently simple two-state system, manipulations of a single environmental driver revealed a rich array of complex temporal dynamics. Nevertheless, with a replicated, controlled system, and the ability to lift proteomic fingerprints from the active microbes, we could successfully identify state changes and characterize the tipping points between them (chapters 12 and 13). However, the precise control over the tipping point that we had hoped to achieve has so far proven elusive. More experimental work with the *Sarracenia* microecosystem or other micro- and macroecosystems is needed if we are going to use models of tipping points and regime shifts to manage and control ecosystems in the face of rapid environmental change.

In this regard, the *Sarracenia* microecosystem provides a scalable model for two other important topics. First, ecologists usually interpret the shift from one state to another and its subsequent return to the initial conditions as a return to the original state.[2] In our experiments with *S. purpurea*, pitchers forced into an anaerobic state with additions of prey or bovine serum albumin eventually recovered to control oxygen concentrations (Northrop et al. 2021). However, the taxonomic diversity mapped from protein profiles differed between the initial and final aerobic states, and between ecosystems that had experienced different rates of enrichment (chapter 13, figure 13.5). Did this result reflect taxonomic redundancy in microbes responsible for ecosystem function and the outcome of unmeasured interspecific interactions among them? Or was our focus on [O_2] too myopic, causing us to miss differences in other important ecosystem functions between the aerobic states observed after the recovery from anaerobic conditions? This distinction mirrors long-standing debates about the relative utility and value of ecosystem restoration and rehabilitation (Callicott et al. 1999).

Second, the time scale of change is crucial for interpreting tipping points and regime shifts. Bestelmeyer et al. (2011) illustrated that what we perceive as a regime shift looks very different to an organism with a very different lifespan. A recent analysis of rates of regime shifts as a function of ecosystem area (Cooper et al. 2020) found that larger ecosystems shift proportionately faster than smaller ones (i.e., that the power-law relationship between rate of change and ecosystem area scales sublinearly; see chapter 1, §1.3). Yet Cooper et al. (2020) only worked with human time scales. Although these may be relevant scales for humans to manage (chapter 11, figure 11.2), they may be less reflective of internal system dynamics (Bestelmeyer et al. 2011). Indeed, the hours-long shift from an aerobic to an anaerobic regime in the *Sarracenia* microecosystem is too fast—in terms of our human abilities to do experimental work—for us to forestall. But for the microbes driving the system, this interval of a few hours spans multiple generations—long enough for natural selection and evolution to occur (Ware et al. 2019). At the same time, inferences about, and management of, tipping points and state changes must account for how human activities are altering the temporal (and spatial) scales of many ecological processes (Rose et al. 2017). We see great potential in using the *Sarracenia* microecosystem to do experiments on the time scales relevant to the plant and the inquilines with different lifespans and life histories that live in concert with it.

In sum, the *Sarracenia* microecosystem has been an ecological gold mine. It has yielded decades of important insights into processes ranging from nutrient stoichiometry and atmospheric deposition, through demography and community assembly, to tipping points and ecosystem function. We hope we have convinced you that there are still many treasures left to uncover in this most excellent of ecological model systems.

Appendices

The Natural History of *Sarracenia* and Its Microecosystem

For more than 20 years, our efforts to study *Sarracenia* across different scales of temporal, spatial, and biological organization have been guided by fundamental knowledge of the taxonomy, systematics, and evolutionary history of *Sarracenia purpurea*, the genus *Sarracenia*, and the family Sarraceniaceae; the autecology, morphology, and life history of *S. purpurea*; its associated organisms and their ecological relationships with one another and with *S. purpurea*, and the place of *S. purpurea* in landscapes both real and figurative (see figure 1.2).

A.1 TAXONOMY AND SYSTEMATICS

Sarracenia purpurea L. is one of 11 species of North American pitcher plants in the genus *Sarracenia* L., which is one of three genera in the Occidental plant family Sarraceniaceae Dumort. (Mellichamp and Case 2009; Naczi 2018). Sarraceniaceae is placed within the Asterid order Ericales (Stevens 2020); it evolved independently of the other two families of carnivorous pitcher plants (Cephalotaceae Dumort. [Oxalidales; Rosids] and Nepenthaceae Dumort. [Caryophyllales; core Eudicots]) (Fleischmann et al. 2018).

All the ≈35 species in the Sarraceniaceae are long-lived, perennial, often evergreen, carnivorous herbs. Their leaves are modified into tubular "pitchers" (Arber 1941), which function as traps that attract, capture, and digest prey (mostly insects and other small arthropods; Ellison and Gotelli 2009). Prey digestion releases nutrients that are absorbed directly by the pitchers. The leaves are arranged in rosettes or emerge along linear, branching rhizomes. The root systems of all Sarraceniaceae are comparatively small and contribute relatively little to the total nutrient budget of the plant (Chapin and Pastor 1995; Butler et al. 2008).

Darlingtonia Torr., *Sarracenia*, and *Heliamphora* Benth. are the three geographically disjunct genera within the Sarraceniaceae (figure A.1). The ancestral, monospecific California cobra lily (*Darlingtonia californica* Torr.) grows in serpentine fens in the Siskyou Mountains of southwestern Oregon and southward

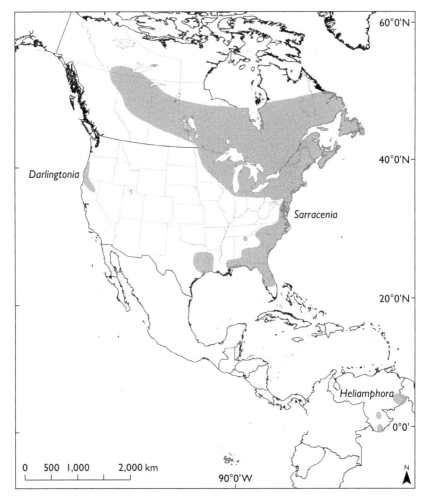

FIGURE A.1. Geographical distribution of the Sarraceniaceae. Map drawn by Robert F. C. Naczi.

into the northern Sierra Nevada Mountains of California. The other two genera in the family diverged from *Darlingtonia* ≈25–44 million years ago (Mya)
(Ellison et al. 2012) and *Sarracenia* diverged from *Heliamphora* ≈14–32 Mya. The
South American sun pitchers (*Heliamphora* spp.) grow on the isolated sandstone
massifs (*tepuis* of the Guayana Highlands of northern South America), whereas
Sarracenia spp. grow in nutrient-poor, well-lit bogs, poor fens, seepage swamps,
and outwash sandplains. *Sarracenia* is the most derived genus in the family and
began to diversify only recently, ≈2–7 Mya (Ellison et al. 2012).

T HVRIS *Limpidi folium aiunt.*

FIGURE A.2. The first known illustration of a *Sarracenia* plant, named *Thuris limpidi* by Pena and de L'Obel (1576) and renamed *Sarracenia flava* by Linneaus. Image courtesy of the Botany Libraries, Harvard University.

A.1.1 Early Descriptions of Sarracenia

A *Sarracenia* species (which appears to be similar to either *S. minor* Walt. or *S. flava* L.) was first named *Thuris limpidi* and illustrated by Pena and de L'Obel (1576, figure A.2). What we now know as *S. purpurea* was first named *Limonium peregrinum* by Clusius (1601, figure A.3), who only observed a dried, flowerless specimen of unknown geographical origin (Cheek and Young 1994). Clusius noted that its leaves had the form of an *Aristolochia* L. (pipevine) flower.[1]

In 1700, Joseph Pitton de Tournefort (1700) renamed Clusius's *Limonium peregrinum* as *Sarracena canadensis*; the name of the genus honored the Canadian botanist Michel Sarrazin de l'Etang, the king's physician in "New France," (the French colonies of North America) who had sent live plants to Tournefort in Paris (figure A.4). Linneaus (1753) retained the genus *Sarracenia* in his *Species Plantarum*, and described two species, *S. purpurea* (the *Limonium peregrinum* of Clusius and the *S. canadensis* of Tournefort) and *S. flava* (the *Thuris limpidi* of Pena and L'Obel).

A.1.2 Biogeography of the Genus

All but one of the 11 species of *Sarracenia* are native to the southeastern United States, where they often grow in small, isolated populations. In contrast, the geographical range of the northern (a.k.a. purple) pitcher plant, *S. purpurea*, is vast (figure A.1). The range of this single species (sensu stricto, thus excluding

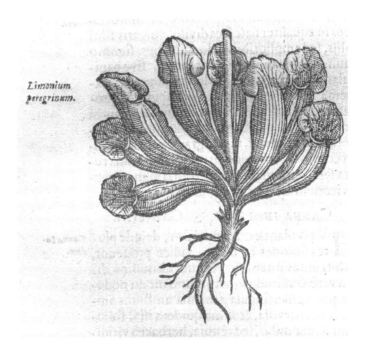

FIGURE A.3. The first known illustration of *Sarracenia purpurea*, named *Limonium peregrinum* by Clusius (1601). Image courtesy of the Botany Libraries, Harvard University.

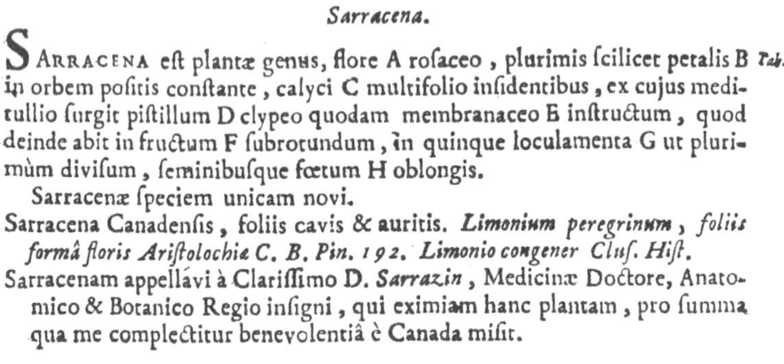

Sarracena.

Sarracena eft plantæ genus, flore A rofaceo, plurimis fcilicet petalis B *Tab. 476.* in orbem pofitis conftante, calyci C multifolio infidentibus, ex cujus medi-tullio furgit piftillum D clypeo quodam membranaceo E inftructum, quod deinde abit in fructum F fubrotundum, in quinque loculamenta G ut pluri-mùm divifum, feminibufque fœtum H oblongis.

Sarracenæ fpeciem unicam novi.

Sarracena Canadenfis, foliis cavis & auritis. *Limonium peregrinum, foliis formâ floris Ariftolochiæ C. B. Pin. 192. Limonio congener Cluf. Hift.*

Sarracenam appellávi à Clariffimo D. *Sarrazin*, Medicinæ Doctore, Anato-mico & Botanico Regio infigni, qui eximiam hanc plantam, pro fumma qua me complectitur benevolentiâ è Canada mifit.

FIGURE A.4. The first taxonomic description of *Sarracenia purpurea* (as *Sarracenia canadensis*). Image from Tournefort (1700) and courtesy of the Botany Libraries, Harvard University.

S. rosea Naczi, Case, & Case) spans 30 degrees of latitude (from northern Florida to Labrador) and 70 degrees of longitude (from Newfoundland to eastern British Columbia; Naczi et al. 1999; Naczi 2018). This remarkable geographical range contributes substantially to our ability to study *S. purpurea* at different spatial scales.

A.1.3 Taxonomic Circumspection

Although all *Sarracenia* species share a common baüplan, there is substantial intra- and interspecific variation in the size, shape, and color of their pitchers and flowers. This variation is the result of the very recent diversification of the genus (Ellison et al. 2012); extensive hybridization between species (Mellichamp and Case 2009; Ellison et al. 2014; Naczi 2018); mutations, selection, and genetic drift (Godt and Hamrick 1996, 1998b; Sheridan and Mills 1998; Sheridan and Karowe 2000); and phenotypic responses to local environmental conditions (Ellison et al. 2004; McPherson and Schnell 2011). Because the early taxonomy was based on pitcher morphology and color, the substantial geographical and morphological variation in *Sarracenia* has been a source of taxonomic confusion. This problem has been compounded because horticulturists have bred, hybridized, and named plant variants for marketing to florists, gardeners, and hobbyists (Ellison et al. 2014; Naczi 2018).

Morphological and molecular data have resolved some of the confusion (Mellichamp and Case 2009; Ellison et al. 2012), but ambiguities remain, especially in the *S. rubra* Walt. and *S. purpurea* species complexes. Within the *S. rubra* complex (McDaniel 1971; Case and Case 1976; Mellichamp and Case 2009; Godt and Hamrick 1998a), taxonomic consensus is further complicated by listings of two subspecies under the Endangered Species Act (US Fish and Wildlife Service 1988, 1989) that are now recognized as distinct species (Mellichamp and Case 2009).

The large geographical range of *S. purpurea* and its substantial morphological variation have led many researchers to identify two subspecies, along with several varieties and forms. Schnell (1979) divided *S. purpurea* into two subspecies, *S. purpurea* ssp. *purpurea* (Raf.) Wherry and *S. purpurea* ssp. *venosa* (Raf.) Wherry. *Sarracenia purpurea* ssp. *purpurea*, the "northern" pitcher plant, grows as far south as Maryland, whereas *S. purpurea* ssp. *venosa*, the "southern" pitcher plant, occurs from Maryland (and possibly New Jersey) south along the East Coast and then across the Florida panhandle, Mississippi, Alabama, and Louisiana. These two subspecies of *S. purpurea* historically co-occurred in southern Maryland and northern Virginia (Wherry 1933), but recent field observations on the degree of sympatry are lacking. Schnell later subdivided *S. purpurea* ssp. *venosa* into three varieties: *S. purpurea* ssp. *venosa* var. *venosa* (Raf.) Fernald, which occurs on the Atlantic coastal plain; *S. purpurea* ssp. *venosa* var. *montana* Schnell & Determann, which occurs in the southern Appalachian Mountains of Georgia and the Carolinas; and *S. purpurea* ssp. *venosa* var. *burkii* Schnell, which grows only on the coastal plain of the Gulf of Mexico from the Florida panhandle westward into Louisiana (Schnell 1993; Schnell and Determann 1997).

Gleason and Cronquist (1991) rejected the subspecific and varietal designations of Schnell (1979) and recognized only two varieties of *S. purpurea*: the "northern" or "purple" pitcher plant, *S. purpurea* var. *purpurea*, and the "southern" pitcher plant, *S. purpurea* var. *venosa*. However, Reveal (1993) pointed out that if Gleason and Cronquist's taxonomy were to be accepted, the correct name for the northern pitcher plant would be *S. purpurea* var. *terrae-novae* de la Pylaie and that of the southern pitcher plant would be *S. purpurea* var. *purpurea*. But Gleason and Cronquist's (1991) proposed lumping has not been accepted in subsequent taxonomic treatments.

Later investigations using isozymes (Godt and Hamrick 1998b) supported the subspecific designations proposed by Schnell (1979), and morphological analysis was used to elevate *S. purpurea* ssp. *venosa* var. *burkii* to species status—as *S. rosea* (Naczi et al. 1999). In their treatment for the Flora of North America, Mellichamp and Case (2009) recognized *S. rosea* as a distinct species and maintained the subspecific identities of *S. purpurea* ssp. *purpurea* and *S. purpurea* ssp. *venosa*, but did not distinguish *S. purpurea* ssp. *venosa* var. *montana* as a separate variety.

Molecular taxonomies of *Sarracenia* based on nuclear DNA have consistently separated *S. rosea* from the rest of the *S. purpurea* complex (Neyland and Merchant 2006; Ellison et al. 2012) with bootstrap support of >99%. However, bootstrap support for further separation of geographical accessions of *S. purpurea* has ranged from only 82% to 88%, suggesting that currently recognized subspecies of *S. purpurea* might reflect poorly sampled clinal variation (Schnell 1978, 1979; Ellison et al. 2004). Analyses of both plastid and mitochondrial DNA supported a separation of *S. rosea* from the rest of the *S. purpurea* complex, whereas analysis of plastid DNA implied that *S. purpurea* ssp. *venosa* var. *montana* is more closely related to *S. oreophila* Wherry than to any other *Sarracenia* species (Ellison et al. 2012). An ancestral hybridization event may have led to *S. purpurea* ssp. *venosa* var. *montana* inheriting its plastid genome via chloroplast capture from sympatric *S. oreophila* or other species in its clade (i.e., *S. alabamensis* Case and R. B. Case). However, more research on natural hybridization in *Sarracenia* is needed to support this hypothesis (Naczi 2018).

A.2 THE BOTANY AND AUTECOLOGY OF *SARRACENIA PURPUREA*

Our generation of community ecologists learned of *Sarracenia purpurea* through Addicott's (1974) classic experimental study of larvae of the pitcher-plant mosquito (*Wyeomia smithii* (Coq.)). Addicott (1974) set out to test whether *W. smithii* acted as a keystone predator (sensu Paine 1966), but he found that it did not increase diversity of lower trophic levels by selectively feeding on dominant competitors. Rather, *W. smithii* reduced the diversity of co-occurring invertebrate prey by filter feeding (see also chapter 10, §10.3.2 and Błędzki and Ellison 1998). In the last half-century, this food web has become a well-studied model system for experimental studies of succession and top-down versus bottom-up controls on food web dynamics (chapters 9 and 10; Miller et al. 2018). But the basic botany and plant (aut)ecology of *S. purpurea* are less familiar to ecologists. We review this rich history of the plant before returning to a description and discussion of its inquiline food web.

A.2.1 Anatomy and Morphology

Plant evolutionary biologists, anatomists, morphologists, and physiologists have focused on the "carnivorous syndrome" (sensu Ellison and Adamec 2018) of *S. purpurea*: its adaptations for attracting, capturing, and digesting animal prey, and for absorbing the released or mineralized nutrients. These adaptations reflect morphological canalization and evolutionary cost–benefit trade-offs between adaptations for prey capture and digestion on the one hand, and physiological requirements of photosynthesis on the other (Givnish et al. 1984, 2018; Ellison and Farnsworth 2005).

Pitchers   The unusual morphology of *Sarracenia* pitchers was the focus of detailed anatomical and morphological study from the late nineteenth through the mid-twentieth centuries. Research by Baillon (1870); Shreve (1906); Troll (1932, 1939); Arber (1941), and Franck (1976) is the basis of currently accepted interpretations of *Sarracenia* leaf morphology. Briefly (figure A.5), the pitchers of *Sarracenia* (and of other carnivorous pitcher plants) are unifacial ("one-sided"; sensu Goebel 1891; Arber 1921), "episcidate" leaves, that is, foliar organs whose ventral (adaxial, or "top") surface is on the inside of a tube (Franck 1976). Whitewoods et al. (2020) presented a new model of regulatory control of the development of pitchers (and carnivorous organs in other, unrelated taxa) based on a simple shift in localization of gene expression in the abaxial and adaxial developmental domains. The leaves emerge from a small underground rhizome, and the plant gradually takes on a rosette shape as new pitchers are produced each year.

Sarracenia pitchers are peltate (i.e., the leaf petiole attaches to the leaf lamina in a position other than the leaf base); in *Sarracenia* the petiole has been modified into a solid rib that runs up and along the back of the pitcher from its bottom to the base of the lid. The pitcher itself is a modified leaf; the broad bottom of the pitcher is homologous to the base of "normal" leaves of other angiosperms. The "keel" is an autapomorphic outgrowth of the ventral side of the leaf, and the "hood" is an autapomorphic extension of the leaf lamina (Franck 1976). In many species of *Sarracenia*, the pitcher body can become much reduced, leaving only an expanded keel manifest as a broad phyllode (chapter 3).

In all species of *Sarracenia* other than those of the *S. purpurea* complex, the hood is reflexed: it covers the "mouth" of the pitcher, preventing rain from filling the pitcher and limiting evaporation of the pitcher's "digestive fluid." In contrast, the hood of *S. purpurea* is upright, allowing rain to fill (and sometimes overflow) the pitcher. Desiccation of *S. purpurea* pitchers declines with pitcher size (Kingsolver 1976). The peristome around the mouth of the pitcher is replete with extrafloral nectaries (Lloyd 1942; Płachno and Muravnik 2018) that secrete sugary compounds to attract prey (primarily ants; Bennett and Ellison 2009; Ellison and Gotelli 2009). Other extrafloral nectaries are present on the colored veins on the outer surface of the pitcher, the pitcher hood, and all over the smooth, upper part of the pitcher (the so-called "attractive" zone; figure A.5).

The inside of the pitcher is divided into three zones. Just below the peristome is a slippery area covered with waxy crystals that make it difficult for insects and other prey to maintain their footing. This region also is covered with many downward-pointing hairs that limit the ability of prey to climb back out of the pitcher (Juniper et al. 1989; Bauer et al. 2018). Below the slippery area, digestive glands cover the inner surface of the pitcher. Finally, the very bottom of the *S. purpurea* pitcher is a smooth, gland-free zone whose function has yet to be determined.

Flowers and Fruits   A mature *S. purpurea* individual produces a single flower on a stalk that extends well above the top of the pitchers (frontispiece). The bud is set at the end of the growing season in late summer and commences growth in early spring before pitchers are produced (Shreve 1906). The downward-pointing flowers have a characteristic "umbrella" shape with five sepals and five drooping petals. What appears as the base of the flower (or the top of the umbrella) is, in fact, an expanded style onto which pollen falls (Shreve 1906; Guo and Halson 2020). Bees and flies are the primary pollinators (Ne'eman et al. 2006), and the umbrella-shaped style also creates a unique chamber for pollinators and a protected surface to retain deposited pollen deposited there, and also presents visual and olfactory cues for pollinators (Guo and Halson 2020). Although *Sarracenia* species are self-compatible, the location of the stigmas at the tips of the upturned style tends to prevent selfing. After the flower is pollinated, the fruit takes the rest of the summer to mature. The fruit is a dehiscent, loculicidal capsule (Shreve 1906) that contains hundreds to thousands of 1–3-mm-long, hydrophobic seeds (Ellison 2001).

A.2.2 Physiology

Two aspects of *Sarracenia* physiology are important for our discussions of scaling of leaf traits (chapters 3 and 4): the digestion and absorption of nutrients from captured prey, and the photosynthetic characteristics of pitchers.

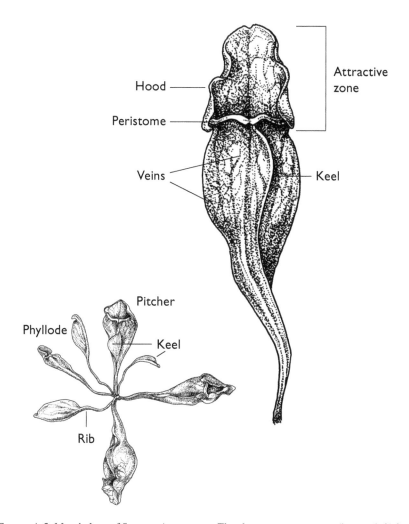

FIGURE A.5. Morphology of *Sarracenia purpurea*. The plant grows as a rosette (bottom left) from a small underground rhizome, and the individual leaves are modified into pitchers or phyllodia. The dorsal "rib" of each leaf is a modified petiole, and the "keel" is an outgrowth of the ventral side of the leaf. The "attractive zone" for prey consists of the vein-rich unreflexed "hood" and the "peristome" surrounding the pitcher's mouth that is replete with extrafloral nectaries. Original drawings by Elizabeth J. Farnsworth.

Digestion and Absorption of Nutrients   Although *Sarracenia* was first described in the seventeenth century, nearly 300 years elapsed before experiments revealed that it actually obtained nutrients from the insects decomposing in its pitchers. Bartram (1793: xii–xiii) noted that all four *Sarracenia* species he observed were "insect catchers," but he doubted that "the insects caught in their leaves, and which dissolve and mix with the fluid, serve for aliment or support to these kinds of plants." Macbride (1818) similarly noted vast quantities of dead insects in *S. flava*, but commented (p. 51) "[w]hat purposes beneficial to the growth of these plants may be effected by the putrid masses of insects, I have never ascertained." In *Insectivorous Plants*, Darwin (1875) presented extensive experimental data on nutrient uptake from decomposed or digested prey by *Drosera* L. and a few other species of carnivorous plants, but he did not discuss *Sarracenia* at all. However, Darwin's notes and letters[2] reveal that although he was interested in determining whether *Sarracenia* could obtain nutrients from captured prey, he failed to obtain convincing data.

Mellichamp (1875) referred to the decomposed insects within pitchers of *Sarracenia variolaris* Michx. (= *S. minor*) as "liquid manure." He gave exquisite descriptions of the natural history of prey capture, pollination, and inquilines of this species, but he did not present any evidence that this liquid manure was actually used by the plant. However, he did note that "after these animal juices have been partially or entirely absorbed by the plant... space. the remaining portion [of the carcasses] commences to dry, until only the backs and legs and shells of the various insects remain."

In the same volume, Canby (1875) described similar mechanisms of prey attraction and capture by *Darlingtonia californica*. The western field naturalist Rebecca Austin also observed that digestive fluid in *D. californica* was produced after the plants had captured insects (cited in Gray 1876) or were fed raw or cooked meat (cited in Ames 1880).

These observations led to numerous experiments in the late nineteenth and early twentieth centuries demonstrating that various species of *Sarracenia*, including *S. purpurea*, could absorb nutrients (reviewed in Hepburn et al. 1927a,b), and that nitrogen was absorbed preferentially relative to phosphorus (Hepburn et al. 1927b, but see Wakefield et al. 2005 and chapter 3). Hepburn and Jones (1927) also documented the presence of proteases in *Sarracenia* spp. and *Darlingtonia*. Hepburn and St. John (1927) demonstrated that new (i.e., unopened) pitchers were sterile, but they found many bacteria in opened pitchers that had already collected rainwater or captured prey. They concluded that bacteria could play a role in prey digestion and mineralization, especially in species that had little (i.e., *S. purpurea*) or no (i.e., *D. californica*) protease expression (Koller-Peroutka et al. 2019). This conclusion would later be supported by observations that *S. purpurea* produces proteases for only a few weeks after pitchers have opened (Gallie and Chang 1997), if at all (Koller-Peroutka et al. 2019), and that bacterial mineralization of prey accounts for virtually all of the nitrogen budget of *S. purpurea* (Butler et al. 2008).

As this book was going to press, Sexton et al. (2020) reported the presence of nitrogen-fixing diazatrophic bacteria in the rhizomes of *S. purpurea*. The importance of nitrogen fixation to its nitrogen budget was not quantified, but Sexton et al. (2020) hypothesized that nitrogen fixation could be adaptive in the nutrient poor habitats where *S. purpurea* grows.

Photosynthesis Carnivorous plants like *S. purpurea* have evolved elaborate traps to catch prey, and most of these traps are modified leaves (Fleischmann et al. 2018). The morphological and structural modifications necessary to attract, capture, digest, and absorb prey are thought to come at the expense of photosynthetic performance (e.g., Givnish et al. 1984, 2018; Ellison and Gotelli 2002; Ellison and Farnsworth 2005). Indeed, photosynthetic rates in carnivorous plants are typically much lower than are found in other perennial herbaceous plants (figure A.6; Ellison 2006). For example, measured mass-based rates of photosynthesis (A_{mass}) in *Sarracenia* rarely exceed 40 nmol $\cdot$ CO_2 g^{-1} $\cdot$ s^{-1} (Farnsworth and Ellison 2008), which is comparable to measured rates in slow-growing evergreen trees and shrubs (figure A.6; Ellison 2006).

A.2.3 Life History

Like all *Sarracenia* species, *S. purpurea* is a long-lived perennial plant. In the field and in cultivation, individuals can live for decades, perhaps even a century or more. Under optimal growing conditions in a greenhouse, plants may flower after 3–5 years, but in the field, most flowering plants that we have seen are at least 10 years old.

Whether an individual flowers in a given year, however, is determined at the end of the previous growing season. Buds of *S. purpurea* and other species of *Sarracenia* are developmentally "preformed": they are set late in the growing season of the year before they develop into flowers (Shreve 1906). It appears that buds are set depending on how much prey (i.e., nutrients) is obtained during the growing season, which in turn is dependent on the number of pitchers a plant produces. Our long-term observations of *S. purpurea* at Hawley Bog in northwestern Massachusetts and Molly Bog in northern Vermont suggest that plants with fewer than six pitchers rarely flower (table A.1).

A further consequence for *S. purpurea* of the developmental preformation of its buds is that flowers emerge early in the growing season, normally before new pitchers are produced (Judd 1959). Early emergence of flowers likely limits the likelihood that pollinators will be captured as prey (see §A.3.3), because when flowers develop and are receptive to pollinators, pitchers are not being produced. However, in the northern part of *S. purpurea*'s geographical range, where the growing season is relatively short, the lag in pitcher production associated with

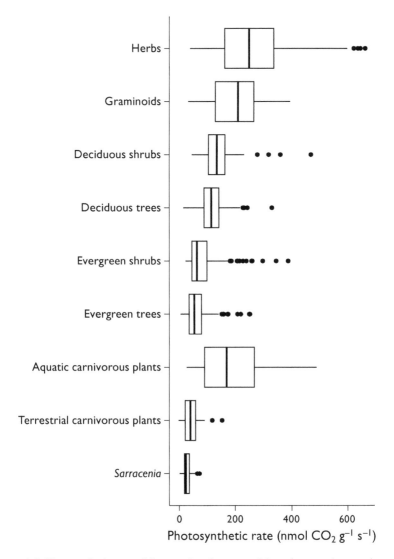

FIGURE A.6. Photosynthetic rates of *Sarracenia*, other terrestrial carnivorous plants, and aquatic carnivorous plants relative to noncarnivorous terrestrial plant species. Data for carnivorous plants from Farnsworth and Ellison (2008) and compilation in Ellison and Adamec (2011); data for noncarnivorous terrestrial plants from the GLOPNET dataset (Wright et al. 2004).

flowering means that a flowering plant rarely produces more than five pitchers (pitchers are generated at roughly two-week intervals). The net result is that it is very unusual for an individual plant to flower two years in a row. We explore the effects on pitcher-plant demography of all of these life-history characteristics in chapters 6 and 7.

TABLE A.1. Reproduction of *S. purpurea* at Hawley and Molly bogs in year $y + 1$ as function of number of leaves produced in year y.

Bog	Number in year y			Number in year $y + 1$		
	Leaves	Plants	Buds	Flowers	Fruits
Hawley	1	1	0	0	0
	2	10	0	0	0
	3	44	1	1	1
	4	87	9	4	4
	5	110	15	10	8
	≥ 6	220	28	19	10
Molly	1	16	0	0	0
	2	50	1	0	0
	3	210	13	11	7
	4	300	18	8	7
	5	228	33	18	18
	≥ 6	494	89	47	28

Computed from data in the Harvard Forest Data Archive dataset HF202.

A.3 ANIMAL–PLANT INTERACTIONS

Sarracenia interacts with animals as a predator of insects (and occasionally small vertebrates), as a host for a detritus-based ("brown") food web of inquilines whose resource base is the carcasses of captured prey, and as prey itself for a small number of specialist and generalist herbivores (figure 1.2; Butler et al. 2008). Inquilines normally are considered to be commensals, but their role in digesting prey, mineralizing nutrients, and oxygenating pitcher fluid imply that they would better be considered mutualists with *S. purpurea* (Bradshaw and Creelman 1984; see also chapter 10, §10.3.3).

A.3.1 Pitcher-Plant Prey

As implied by the title of Darwin's *Insectivorous Plants* (1875), insects account for most of the prey of pitcher plants and other terrestrial carnivorous plants (Ellison and Gotelli 2009). Some of the earliest and most detailed studies of prey capture in *Sarracenia* were done in the early twentieth century by Frank Morton Jones (summarized in Jones 1935).[3] Other comprehensive studies of *Sarracenia* prey capture include Wray and Brimley (1943); Judd (1959), and Cresswell (1991); dissertations by Gibson (1983) and Folkerts (1992); and in situ videography by Newell and Nastase (1998).

All *Sarracenia* species are generalist insect predators, but ants are their most common prey (table A.2). Their diet also may include beetles, spiders, slugs, other arthropods, and, on rare occasions, even small reptiles and amphibians (Jones 1935; Wray and Brimley 1943; Judd 1959; Gibson 1983; Cresswell 1991; Folkerts 1992; Heard 1998; Newell and Nastase 1998; Schnell 2002; Butler et al. 2005; Bennett and Ellison 2009; Moldowan et al. 2019). However, the relative abundance of different species of prey is proportional to their relative abundance in the local habitat, which can vary among bogs. Prey relative abundance does not reflect prey specialization by the genus *Sarracenia* (Ellison and Gotelli 2009), although at least one population of *S. purpurea* preys disproportionately on spiders (Milne and Waller 2018). There also is no evidence for niche adjustment and dietary specialization among co-occurring *Sarracenia* species; in fact, there is usually more dietary overlap between *Sarracenia* species than expected by chance (table A.3; Ellison and Gotelli 2009).

Even though pitchers are often replete with carcasses of ants and other prey, video observations of prey capture by *S. purpurea* revealed that it is a relatively inefficient predator: <2% of ants visiting the plant actually are captured (Newell and Nastase 1998). Other pitcher plants, including *Darlingtonia californica* (Dixon et al. 2005) and *Nepenthes rafflesiana* Jack. (Bohn and Federle 2004; Bauer et al. 2008) also capture prey at low rates, at least in sunny and dry conditions. However, when the pitcher lip (peristome) of *N. rafflesiana* is wetted by rain, condensation, or nectar secreted by the extrafloral nectaries that line the peristome, it becomes nearly frictionless. Under these conditions, nearly all foraging ants then slip off the peristome and into the pitcher (Bauer et al. 2008).

As in *Nepenthes*, all *Sarracenia* have extrafloral nectaries around the peristome (Vogel 1998; Płachno et al. 2007). Two centuries ago, Macbride (1818: 52) described the peristome of *S. flava* as nearly frictionless: "[the peristome] is either covered with an impalpable and loose powder, or that the extremely attenuate pubescence is loose. This surface gives to the touch the sensation of the most perfect smoothness." The wettable properties of the peristome of *S. purpurea* have not been investigated, but they are probably similar to those of Asian pitcher plants (Bauer et al. 2018).

Pitchers of *S. purpurea* are most effective at capturing prey during the first month or so after they open (Fish and Hall 1978; Wolfe 1981; Cresswell 1991), and new pitchers are produced every 10 to 14 days. Nutrients from prey captured by the first pitchers of the season are used to produce subsequent pitchers (Butler and Ellison 2007). Nutrients remaining in senescent leaves at the end of the growing season are stored in the overwintering, evergreen leaves and in the rhizome. These stored nutrients are mobilized in the following spring to produce the first pitchers of the year (Butler and Ellison 2007).

TABLE A.2. Prey spectra of *Sarracenia* species.

	S. alata	*S. flava*	*S. leucophylla*	*S. minor*	*S. psittacina*	*S. purpurea*	*S. rubra*
Formicidae	66	60	24	98	42	40	71
Diptera	11	6	32	2	2	28	15
Coleoptera	6	19	10	0	6	11	1
Araneae	0	0	0	0	2	3	1
Other Hymenoptera	5	3	24	1	0	2	8
Lepidoptera	2	3	6	0	2	2	1
Homoptera	1	3	0	0	0	2	0
Orthoptera	0	0	0	0	0	2	0
Hemiptera	2	1	1	0	0	0	1
Acarina	1	0	0	0	0	0	0
Trichoptera	1	0	0	0	0	0	0

Values are median percentage of each arthropod order (or ants [Formicidae]) collected from $n = 1$–10 individuals of each species. Data from the Harvard Forest Data Archive dataset HF111 and compiled by Ellison and Gotelli (2009).

TABLE A.3. Niche overlap in prey use by congeneric *Sarracenia* spp.

Location	Species	Niche overlap Observed	Expected	P
Okaloosa County, Florida	5[a]	0.64	0.20	0.002
Santa Rosa County, Florida	2[b]	0.99	0.13	0.038
Turner County, Georgia	3[c]	0.63	0.24	0.013
Brunswick County, North Carolina	3[d]	0.98	0.13	0.001

Each row is from a different study, and gives the number of co-occurring species, the observed average pairwise niche overlap (Pianka 1973) based on prey composition, the expected mean value of pairwise niche overlap based on 1000 randomizations of the prey capture data, and the *P* value of the upper tail probability of finding the observed value if the data had been drawn from the null distribution. Based on data from Folkerts (1992) and analysis in Ellison and Gotelli (2009).
[a] *S. flava, S. leucophylla, S. rubra, S. purpurea, S. psittacina.*
[b] *S. flava, S. psittacina.*
[c] *S. flava, S. minor, S. psittacina.*
[d] *S. flava, S. purpurea, S. rubra.*

A.3.2 The Inquiline Food Web

The pitchers of *S. purpurea* fill with rainwater or snow melt; the hoods of the other species cover the pitcher opening and digestive fluid produced by the plant fills their pitchers. All pitcher plants that form a liquid-filled trap also have an associated inquiline food web that develops in this specialized habitat. Food webs of other Sarraceniaceae have been poorly studied (but see Naeem 1988 and Nielsen 1990 for *Darlingtonia californica*) and have fewer species than that of *S. purpurea*. We discuss the organization of the *S. purpurea* food web in great detail in chapters 9 and 10; here we describe its general taxonomic diversity.

Captured prey are the resource base of the *S. purpurea* food web (Addicott 1974; Butler et al. 2008; Miller et al. 2018; figure A.7). The drowned prey are first shredded by larvae of the chironomid midge, *Metriocnemus knabi* Coq. and early instars of the sarcophagid fly *Fletcherimyia fletcheri* (Aldrich). The shredded detritus is then decomposed further by bacteria. These in turn are fed on by protozoa, bdelloid rotifers (especially *Habrotrocha rosa* Donner), and larvae of the pitcher-plant mosquito, *Wyeomyia smithii* (Addicott 1974; Heard 1994). Other prominent members of this food web include the histiostomatid mite *Sarraceniopus gibsoni* (Nesbitt) and later instars of *F. fletcheri*, which also are cannibalistic.

There are relatively few taxa of macroinvertebrates and protozoa in the food web of an individual *S. purpurea* leaf (figure A.7). Although a few protist genera

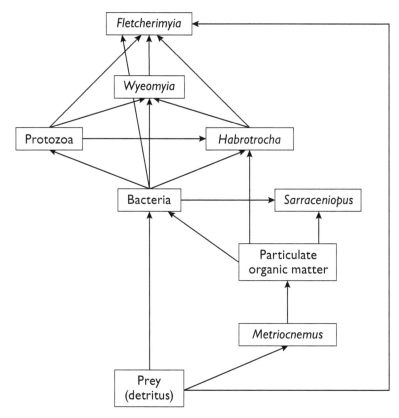

FIGURE A.7. Main components of the *Sarracenia purpurea* food web. Captured prey is shredded by *Metriocnemus knabi* and *Fletcherimyia fletcheri* larvae into particulate organic matter (POM) and directly decomposed by bacteria. *Sarraceniopus gibsoni* and *Habrotrocha rosa* feed on POM. Bacteria are consumed by protozoa and *H. rosa*, and all three are preyed upon by the top predators in the system: larvae of *Wyeomyia smithii* and first- and second-instar *W. smithii* larvae.

(*Bodo* Ehrenberg, *Colpidium* Stein, *Colpoda* O. F. Müller) and the dominant macroinvertebrates (*F. fletcheri, W. smithii, M. knabi, S. gibsoni, Habrotrocha rosa*) occur in most pitchers sampled throughout the range of *S. purpurea* (Judd 1959; Błędzki and Ellison 1998; Buckley et al. 2003, 2010; Baiser et al. 2012; Parain et al. 2019), many more taxa occur less frequently. Addicott (1974) listed 16 flagellates and at least 23 ciliates collected from pitchers in three bogs near the University of Michigan Biological Station. Błędzki and Ellison (2003) collected six species of rotifers from 31 bogs in Massachusetts, Vermont, and Connecticut.

Jones (1935) noted the occurrence of spiders that spin webs over the mouths of *Sarracenia*, and Milne (2012) discussed the use of older, senescent *S. purpurea*

TABLE A.4. Ants found associated with *S. purpurea* in the six New England states: Maine, New Hampshire, Vermont, Massachusetts, Connecticut, and Rhode Island.

	ME	*NH*	*VT*	*MA*	*CT*	*RI*
Ponera pennsylvanica Buckley						✓
Dolichoderus mariae Forel	✓					
Dolichoderus pustulatus Mayr	✓	✓	✓	✓	✓	✓
Tapinoma sessile (Say)	✓	✓	✓	✓	✓	✓
Camponotus herculeanus (L.)	✓		✓			
Camponotus nearcticus Emery				✓		
Camponotus novaeboracensis (Fitch)				✓		
Formica argentea Wheeler	✓					
Formica aserva Forel	✓					
Formica neorufibarbis Emery	✓					
Formica subaenescens Emery	✓	✓	✓	✓	✓	✓
Formica subsericea Say	✓					
Lasius aphidicolus (Walsh)				✓		
Lasius americanus Emery	✓					✓
Lasius neoniger Emery	✓					
Crematogaster cerasi (Fitch)					✓	
Leptothorax canadensis Provancher	✓					
Myrmica fracticornis Forel	✓		✓			
Myrmica incompleta Provancher	✓					
Myrmica latifrons Stärcke	✓					
Myrmica lobifrons Pergande	✓	✓	✓	✓	✓	

Compilation from data in the Harvard Forest Data Archive, dataset HF159.

pitchers as oviposition sites for five spider species in Virginia and North Carolina. Throughout New England, the ant *Dolichoderus pustulatus* Lund builds carton nests in dead or dying leaves of *S. purpurea* (A. Ellison and N. Gotelli, *field observations*; see lower right inset of figure 1.2). In the same region, other ant species, including *Myrmica lobifrons* Pergande, *M. alaskensis* Wheeler, and *Formica subaenescens* Emery, nest around the base of *S. purpurea* rosettes, but this association more likely reflects local environmental conditions rather than a host-specific association (table A.4).

Fletcherimyia fletcheri, *Wyeomyia smithii*, and *Metriocnemus knabi* are known only from *S. purpurea*, but larvae of nine other sarcophagid fly species (in the genera *Sarcophaga* Meigen and *Fletcherimyia* Townsend) have been collected from southeastern *Sarracenia* species (Dahlem and Naczi 2006). Another species of midge, *Metriocnemus edwardsi* Jones, feeds on captured prey in *Darlingtonia californica* (Naeem 1988). The chloropodid flies *Aphanotrigonum darlingtoniae* (Jones) and *Tricimba wheeleri* Mlynarek have been described, respectively, as inquilines in *Darlingtonia californica* and in two species of *Sarracenia* in the southeast United States (Jones 1916, 1920; Folkerts 1999; Mlynarek and Wheeler 2018).

Mites in the genus *Sarraceniopus* Fashing & OConnor occur only in Sarraceniaceae pitchers, and they have been collected from *Darlingtonia californica* (the eponymous *Sarraceniopus darlingtoniae* Fashing & OConnor; Fashing and OConnor 1984), all species of *Sarracenia*, and several species of *Heliamphora* (R.F.C. Naczi, *personal communciation*).

In contrast to the modest species diversity of inquiline macroinvertebrates, the diversity of microbes in the pitchers of *S. purpurea* is very high. Only a handful have been successfully cultured (e.g., Cochran-Stafira and von Ende 1998; Miller and terHorst 2012; Zhang et al. 2020a,b), but molecular probes have revealed large numbers of microbial operational taxonomic units (OTUs). For example, Peterson et al. (2008) sequenced and analyzed T-RFLPs of 16S rRNA from three New England bogs. They identified 133 gene fragments, mostly representing rare and unidentifiable microbial taxa, but the few common and identifiable OTUs included *Craurococcus roseus* Saitoh et al. and the enteric bacteria *Citrobacter freundii* (Braak) Werkman & Gillen, *Hafnia alvei* Møller, *Pantoea agglomerans* (Ewing and Fife) Gavini et al., *Rahnella aquatlis* Izard et al., *Serratia fonticola* Gavini et al., *S. plymuthica* (Lehmann and Neumann) Bergey et al., and an unidentified *Serratia* Bizio. Paise et al. (2014) and Canter et al. (2018) assayed community structure of Archaea and Bacteria from *S. rosea* growing in the Apalachicola National Forest in Florida using iTag sequencing of 16S rRNA genes. These communities included >1000 OTUs, dominated by gene fragments assignable to the phyla Proteobacteria and Bacteroidetes. Peterson et al. (2008) and Boynton et al. (2019) cultured various yeasts, including *Candida pseudoglaebosa* M. Suzuki & Nakase, *Moesziomyces aphidis* (Henninger & Windisch) Q.M. Wang, Begerow, F.Y. Bai & Boekhout, and *Rhodotorula babjevae* (Golubev) Q.M. Wang, F.Y. Bai, M. Groenew. & Boekhout from *S. purpurea*.

The bacterial assemblages in other species of *Sarracenia* other than *S. purpurea* are easily as diverse (Koopman et al. 2010; Koopman and Carstens 2011; Satler et al. 2016; Sexton et al 2020), and the food web of *Darlingtonia californica* similarly includes flagellates, ciliates, and hundreds of bacterial OTUs (Armitage 2016a,b, 2017).

A.3.3 Herbivores and Pollinators

The herbivores and pollinators of *S. purpurea* have been studied much less than the inquiline food web. They are important to understand because there should be strong selection for carnivorous plants to evolve effective mechanisms not only to capture prey, but also to avoid trapping their pollinators (Givnish 1989; Ellison and Gotelli 2001; Cross et al. 2018). On the other hand, herbivores need to avoid being trapped as prey while grazing the plant.

Herbivores   Moth larvae are among the few organisms that feed on *Sarracenia* tissue (Rymal and Folkerts 1982; Lamb and Kalies 2020). *Papaipema appassionata* (Harvey) is a specialist root-borer that kills entire plants. Its larvae bore into the rhizomes of both *S. purpurea* and *S. flava*, killing entire *S. purpurea* plants or large clumps of *S. flava* (Jones 1908; Atwater et al. 2006). As the larva feeds, a characteristic pile of frass accumulates at the center of the *S. purpurea* rosette. The plant wilts, and the pitchers turn yellow and then black as the plant dies within a single field season (Atwater et al. 2006). Larvae pupate in late summer at the base of the plant and adults emerge that same summer and oviposit single eggs on other plants (Bird 1903; Brower and Brower 1970). *Papaipema appassionata* is patchy and rare in New England, but it can be an important source of mortality when it appears in a population.

Three species of the leaf-mining genus *Exyra* Grote feed on *Sarracenia* leaves. *Exyra fax* Grote feeds on *S. purpurea* throughout its range; *E. ridingsii* Riley feeds on *S. flava* and occasionally *S. minor*; and *E. semicrocea* Guenée, which has the broadest range of host plants, feeds on *S. alata* (Wood) Wood, *S. leucophylla* Raf., *S. minor*, *S. psittacina* Michx., and *S. rubra* (Jones 1921; Folkerts and Folkerts 1996). The adults of *E. ridingsii* and *E. semicrocea* lay a single egg on the inner wall of a pitcher, whereas *E. fax* lays multiple eggs in a *S. purpurea* pitcher. Newly hatched larvae *E. fax* disperse among the pitchers of a single rosette so that only one larva occurs in each pitcher (Jones 1921). Each larva then girdles or chews a drainage hole at the base of its pitcher. The damaged leaf cannot capture prey or, in the case of *S. purpurea*, host a food web, but *Exyra* does not kill the plant outright (Fish 1976; Rymal and Folkerts 1982; Folkerts and Folkerts 1996). After draining the pitcher, the larva seals the top of the leaf either with a web of silken threads or by girdling and closing the leaf just under the pitcher lip. The larva then feeds on the interior surface of the pitcher wall (Jones 1921; Rymal and Folkerts 1982; Folkerts and Folkerts 1996). Larvae overwinter as late instars, emerge in the early spring to feed again on flowers or young pitchers, and then pupate in the leaf tissue.

The larvae of the olethreutid moth *Endothenia hebesana* (Walker) feed on the developing carpels of *S. purpurea* flowers (Jones 1908; Hilton 1982). Adults lay single eggs on flower bracts, and the first instars burrow into the base of the ovary, where they feed on the seeds either from within (first instars) or without (second–fifth instars). Larvae then pupate and overwinter in the flower stalks. Adults emerge in early spring, approximately when the plants are flowering. Hilton (1982) mentions two ichneumonid parasitoids of *E. hebesana* larvae and a *Trichogramma*

Riley species that parasitizes its eggs, but no detailed studies of either the moth or its parasitoids have been published.

The prostigmatid mite *Leipothrix darlingtoniae* Fashing was described from individuals collected from above the fluid level of the inner walls of *Darlingonia californica* pitchers (Fashing 1994). This genus of mites includes a number of plant parasites and gall formers, some of which are considered pests; others are being investigated as biological control agents for nonnative plants. The mite's impact on *Darlingtonia* has not been studied, nor have other members of the genus been reported from other Sarraceniaceae.

Pollination   Like virtually all other carnivorous plants (Cross et al. 2018), *S. purpurea* is pollinated by insects (Jones 1935; Ne'eman et al. 2006). Although all *Sarracenia* species are self-compatible, self-pollination is rare (Mandossian 1965; Burr 1979; Thomas and Cameron 1986). Inbreeding depression has been observed in *S. flava* (Sheridan and Karowe 2000) and is associated with anthocyanin-free mutants in many *Sarracenia* species, including *S. purpurea* (Sheridan and Mills 1998). Agamospermy (parthenogenesis), however, has not been reported: an insect vector is required to either self-pollinate or outcross the flowers.

The umbrella-shaped flowers (frontispiece) present a curtain of downward-hanging petals and upward-pointing stylar lobes. Pollinators must pass through this curtain to obtain nectar and reach the anthers and stigmas. Large bumblebees (e.g., *Bombus vagans* Smith, *B. bimaculatus* Cresson) and adults of *Fletcherimyia fletcheri*, the sarcophagid fly whose larval stage is the top predator in the inquiline food web of *S. purpurea* (Ne'eman et al. 2006), are the most commonly observed pollinators.

Sarracenia purpurea uses both spatial and temporal separation of flowers and traps to minimize prey–pollinator conflict. *Sarracenia purpurea* not only flowers before new pitchers are produced (§A.2.3); it also holds its flowers well above the tops of older pitchers (frontispiece). Both of these traits minimize the chances for accidental capture of pollinators in the trap.

A.4 BIOGEOGRAPHY AND LANDSCAPE ECOLOGY

Sarracenia purpurea is a classic "sparse" species (sensu Rabinowitz 1981). Although it grows in a very restricted range of environmental conditions— ombrotrophic ("rain-fed") bogs, poor fens, nutrient-poor streamside wetlands, coastal sandplains, and seepage swamps—individual populations can be large and dense. In the states along the southern and western edges of its native range in the United States, *S. purpurea* is considered a species of conservation concern. In the northern and eastern states, and throughout Canada, the bogs and poor fens in which *S. purpurea* grows often are protected, and its populations are considered secure. Carnivorous plant enthusiasts have introduced *S. purpurea* into the western United States and throughout Europe. In many European localities, it has become naturalized and is considered either a species of management concern or an invasive species (§A.4.2).

A.4.1 Continental Biogeography of Sarracenia purpurea *and Its Inquilines*

Among the 11 species of *Sarracenia, S. purpurea* has by far the largest geographical range, and is the only species in the genus found outside of the southeastern United States. Its range—from Newfoundland and Labrador west across Canada to eastern British Columbia, south along the eastern seaboard of the United States to Florida and Mississippi, and west within the United States into Ohio, Illinois, Iowa, and northern Indiana (figure A.8)—encompasses $> 3.5 \times 10^6$ km^2 (Naczi et al. 1999; Naczi 2018). Remarkably, the structure and composition of the macroinvertebrates of the inquiline food web of *S. purpurea* is very consistent across this entire geographical range (figure A.8; Buckley et al. 2003, 2010; Baiser et al. 2012). This constancy is in marked contrast with that observed for other insect–plant associations, which usually vary in species composition across large geographical ranges of a host plant (Agosta 2006). However, the microbiomes of *S. purpurea* and other *Sarracenia* species are geographically less consistent and tend to be more similar within than between bogs (Peterson et al. 2008; Gray et al. 2012).

Throughout this broad range, however, *S. purpurea* and its obligate inquilines are found only in isolated populations because their required habitats—bogs, poor fens, seepage swamps, and nutrient-poor outwash plains—occur very patchily across the landscape. Habitat suitability models (a.k.a. species distribution models, environmental niche models, and climate-envelope models; Guisan et al. 2017) are a popular tool for forecasting future geographical range shifts. But most of these models implicitly assume that species can move unimpeded across geographical space in their tracking of climatic change. For genera like *Sarracenia*, this assumption does not hold because the matrix of "natural" habitats (forests and fields) and expanding areas of urbanized and planted land are unsuitable either for establishment or for stopovers by long-distance dispersers (Fitzpatrick and Ellison 2018).

A.4.2 Dispersal, Establishment, and Invasion

For *Sarracenia* (and its associated inquilines), a significant barrier to successful long-distance dispersal is that its habitat is rare and patchily distributed (Croizat

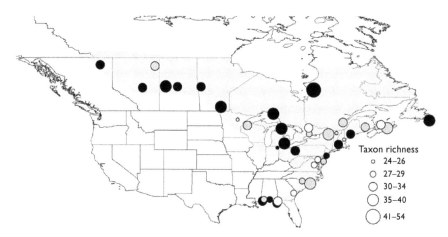

FIGURE A.8. Map showing the locations of 39 sites (circles) sampled by Buckley et al. (2010) across the geographical range of *Sarracenia purpurea* (shaded area) in North America. The size of each point represents the number of inquiline taxa occurring within pitchers at that site out of the 90 encountered across all sites; larger points represent sites with a greater number of taxa. Percentage similarity in species composition of all taxa is represented by the shading of the points: sites with <70% similarity are white; those between 70% and 80% similarity are gray, and those with >80% are black. Reproduced with permission of the authors and John Wiley & Sons from figure 1 of Buckley et al. (2010).

1952). Throughout the northern part of its range, this habitat has been available only since the end of the most recent glaciation (ca. 14–11 kya).

Since the end of the nineteenth century, plant biologists have been perplexed by the paradoxical observation that many plant species that successfully recolonized northern latitudes following the last glacial recession, disperse poorly on average, and lack obvious adaptations for effective long-distance dispersal (Reid 1899). The eponymous "Reid's paradox" is ecologically important because plants will need to disperse and successfully colonize new habitat patches rapidly if they are to keep pace with ongoing climatic change and habitat fragmentation caused by anthropogenic activities.

For trees with generalist habitat requirements, Reid's paradox largely has been resolved (Clark et al. 1998). Although the average dispersal distances of many tree species may be short, the distribution of dispersal distances is "fat-tailed" and occasional dispersal events of very long distance have allowed tree species to shift their ranges very effectively in response to rapid changes in past climates (e.g., Clark 1998; Higgins and Richardson 1999). But it is unclear whether this mechanism applies to herbs, which typically have very short dispersal distances (Cain et al. 2000).

Sarracenia seeds are very small, and they have no obvious ornamentation, elaiosomes, or other structures to attract potential long-distance dispersers. Long-distance dispersal of the hygroscopic *Sarracenia* seeds could occur by flotation, but even then, the maximum dispersal distance recorded for water-dispersed herbs rarely exceeds 4 m (Waser et al. 1982). If *S. purpurea* seeds have a similar maximum dispersal distance, or even one that is one or two orders of magnitude greater, they are still unlikely to disperse easily between bogs separated by tens to hundreds of kilometers. Thus, only very rare, long-distance dispersal events and subsequent successful establishment can explain the broad range expansion of *S. purpurea* since the end of the Pleistocene glaciation. Given the rapid pace of contemporary climatic change, it is uncertain that waterborne dispersal will allow for successful geographical range shifts and prevent the extinction of *S. purpurea* populations over the next century Fitzpatrick and Ellison (2018).

Within a site, *S. purpurea* seedlings are observed commonly to cluster together near the parent plant. Ellison and Parker (2002) studied seed dispersal of *S. purpurea* at Hawley Bog. They arrayed sticky seed traps in a regular pattern around five spatially isolated focal plants, each of which bore a single maturing seed capsule. Seed dehiscence occurred in mid-October, and dispersal continued throughout the fall. The seed-dispersal curve was relatively long tailed: the mean dispersal distance was 12.8 cm from the parent plant, but the median dispersal distance was only 5 cm from the parent plant (figure A.9). These experimental results were mirrored by the distribution of all *S. purpurea* individuals recorded growing in two 5 × 5 m plots at Hawley Bog. In both plots, juvenile and adult *S. purpurea* plants were spatially clustered (Ellison and Parker 2002).

We estimated the establishment probability of seeds from grids of 100 seeds planted just below the *Sphagnum* surface adjacent to our dispersal traps. We planted 500 seeds in the grids in Hawley Bog, but we recorded only 20 germinants after two years of monitoring (Ellison and Parker 2002). Combined with observations of fruit maturation and predation by *Endothenia daeckeana*, and early seedling survival, we estimated that a plant flowering in one year would produce on average four established seedlings the following year. This estimate was used in our demographic modeling of *S. purpurea* (chapter 6). Simulations of dispersal and establishment of *S. purpurea* using our estimated parameters (median dispersal distance = 12.8 cm, 4% establishment probability) yielded spatial distributions of adult plants (expressed as distance-to-nearest-neighbors) that did not differ from field observations (Ellison and Parker 2002).

Despite its short average dispersal distance, the population genetic structure of *S. purpurea* suggests that long-distance dispersal events followed by establishment and rapid population growth occasionally have occurred, and may have led to population diversification by isolation. The infraspecific taxa of *S. purpurea* (§A.1.3)

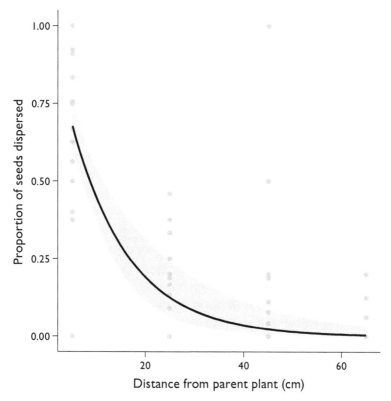

FIGURE A.9. Seed dispersal of *Sarracenia* (after Ellison and Parker 2002). Gray circles are raw data from all seed traps and the line is the best-fit negative exponential curve (with 95% confidence region).

are distinguished morphologically (e.g., Schnell 1979, 1993; Schnell and Determann 1997; Naczi et al. 1999; Ellison et al. 2004; Mellichamp and Case 2009), to a lesser extent by seed size and dormancy requirements (Ellison 2001), and by isozymes (Godt and Hamrick 1998b) and nuclear, plastid, and mitochondrial DNA (Ellison et al. 2012). Rare long-distance dispersal events could have resulted in the distributional pattern of infraspecific taxa of *S. purpurea* if diversification occurred in peripheral bogs following dispersal.

Alternatively, far-north populations of *S. purpurea* could have recolonized from nearby *nunataks* (unglaciated mountaintops) following glacial retreat (Croizat 1952). However, current evidence suggests that the closest ones were in Greenland, not in eastern North America (Bierman et al. 2015). Either way, measurements of finer-scale genetic differentiation of individual populations could help to resolve the importance of founder effects in *S. purpurea*, especially in marginal populations within the native range of *S. purpurea* and in isolated populations where it has been introduced (Schwaegerle and Schaal 1979).

For example, in the village of Buckeye Lake in Licking County, Ohio, several chunks of *Sphagnum* moss broke loose from the "Big Swamp" when the current Buckeye Lake was created in 1830 by blocking the swamp's drainage stream during the construction of the Erie Canal. One of the larger islands is now known as the Cranberry Bog State Nature Preserve. In 1912, a single *S. purpurea* individual was planted onto this floating bog by Freda Detmers, a graduate student in botany at Ohio State University. Ten years later, there were hundreds of pitcher plants, and by the late 1970s, Kent Schwaegerle, another graduate student at Ohio State, estimated that there were more than 150,000 pitcher plants covering ≈1.2 ha of this little bog (Schwaegerle 1983).

Sarracenia purpurea grows elsewhere in Ohio, so it is not considered a weed or an invasive species in Cranberry Bog. This is not the case in other parts of the world (e.g., Parisod et al. 2005; Adlassnig et al. 2010). Along with other carnivorous plants, *S. purpurea* has been introduced into bogs and fens by well-meaning scientists, horticulturalists, and avid plant collectors. Like the fictional triffids (Whyndham 1951), populations of carnivorous plants grow rapidly, occasionally escape from cultivation, and can become noxious, threatening weeds—or new islands of biodiversity.

In Switzerland, *S. purpurea* were introduced by botanists in the late nineteenth century. In the "Les Tenasses" peat bog, *S. purpurea* was planted around 1919; less than 100 years later, they were growing at a density of more than 40,000 plants per acre (≈16,000 plants/ha) (Feldmeyer 1985), about the same density as in Ohio's Cranberry Bog. At Les Tenasses, pitcher plants are considered a "véritable mauvaise herbe" ("truly bad plant") that are overrunning and outcompeting rare native plant species (Parisod et al. 2005; Adlassnig et al. 2010).

Sarracenia purpurea was introduced into Ireland in the early 1900s (Walker 2014). Populations there, and elsewhere in the United Kingdom, are quite dense, and can account for >50% of the vegetation cover in some peatlands (Foss and O'Connell 1985; Couwenberg and Joosten 2004). Since 2015, *S. purpurea* has been considered to be invasive in the United Kingdom (Cox et al. 2015), and eradication efforts are now underway (Walker et al. 2016). *Sarracenia purpurea* also has been introduced successfully in France, Germany, and Denmark, and flowering populations of *S. flava* are known from the Alsace, Franche-Comté, and Haute-Savoie regions of eastern France (Adlassnig et al. 2010). None of these introductions are as yet classified as a management concern, although in many European countries, *Sarracenia* individuals outplanted by carnivorous plant enthusiasts are removed as soon as they are found (Adlassnig et al. 2010).

Last, introduced pitcher plants (*S. purpurea* and *S. flava*) are being watched closely in New Zealand (Hicks et al. 2001; Heenan et al. 2004). There, Cape sundews (*Drosera capensis* L.) and creeping bladderworts (*Utricularia livida* E. Mey.) already are considered modern-day triffids: "carnivorous weeds on the loose" (New Zealand Department of Conservation, n.d.).

The Basics of Resource Limitation

Plants growth and reproduction are affected by two kinds of limiting resources. The first limits plant growth and reproduction when its absolute concentration is too low for physiological processes to occur. The second limits plant growth and reproduction depending on relative concentrations—stoichiometric ratios between individual nutrients (Gusewell 2004).

B.1 ABSOLUTE RESOURCE LIMITATION

For plants, light (Schmitt and Wulff 1993), water (Schenk and Jackson 2002), and CO_2 (Rogers et al. 1994) need to be available above a minimum amount because they play central roles in photosynthesis and carbon fixation (Schimel 1995). All other things being equal, plant growth responds to changes in the absolute concentration or availability of CO_2, water, or light. Insufficient light changes a plant's growth form (e.g., etiolation), slows individual growth, diminishes reproductive effort or success, and increases mortality rates. Thus, plants have evolved a variety of adaptations to efficiently capture and use light and to compete with neighboring plants for access to light (Givnish 1979, 1986).

Without sufficient water, plants wilt and die. Adaptations to water limitation are associated with some of the most reliable morphological (e.g., wax-coated or reduced leaves), physiological (e.g., stomatal conductance), and photosynthetic (C3, C4, CAM pathways) differences among plant species (Díaz et al. 2016). CO_2 also is essential for photosynthesis. Its concentration in the atmosphere varied relatively little throughout the Holocene. However, since the onset of the Industrial Revolution in the late nineteenth century (Petit et al. 1999), deforestation and widespread burning of fossil fuels has increased the atmospheric $[CO_2]$ from 280 parts per million (ppm) to its current levels of >400 ppm, the highest they have been in the past 800,000 years (chapter 2, figure 2.3, and Foster et al. 2017; NOAA 2018). In the absence of water or nutrient limitation, this increase in atmospheric $[CO_2]$ can lead to increases in photosynthetic rate and growth (the "CO_2 fertilization effect"; Huang et al. 2007).

B.2 RELATIVE RESOURCE LIMITATION AND THE SPRENGER–LIEBIG LAW OF THE MINIMUM

Without sufficient nutrients, photosystems fail, biochemical cascades grind to a halt, and plants starve to death. For many years, studies of nutrient limitation, like studies of light limitation, focused on absolute quantities of available nutrients. However, farmers have long known that crop growth could be enhanced not by adding any or all nutrients, but rather by adding particular nutrients. This observation, formalized in the Sprengel–Liebig Law of the Minimum (Sprengel 1828; Liebig 1840; van der Ploeg et al. 1999) is the principle that plant growth is controlled by a single limiting factor—the scarcest resource available (Tilman 1982). Because crop growth is limited most often by N, P, or K, a great deal of effort has been invested in developing methods to manufacture or extract NO_3, NH_4, and NH_4NO_3 (e.g., the Haber-Bosch process for artificial nitrogen fixation from gaseous N_2; Razon 2018); PO_4 (e.g., acidification of apatite from phosphate rock; Andersson et al. 2016); and potash (mining and evaporation of brine; Talbot et al. 2009). In parallel, ecological studies of relative nutrient concentrations (stoichiometric ratios) have focused on these same three macronutrients (e.g., Chapin et al. 1986; Ågren 2008).

Plant ecologists are most familiar with N and P stoichiometry, for which the optimal stoichiometric ratio is $\approx 15:1$ (N:P). If N is added when N:P is near-optimal, the N:P ratio increases. Phosphorus, not N, then becomes limiting, and nutrient deficiencies will occur even if the absolute concentrations of P would be considered sufficient for optimal plant growth (Ågren 2008). Physiological and morphological mechanisms for maintaining optimal stoichiometric ratios in natural conditions are likely to be key evolutionary adaptations that reflect strong selective pressures. Environmental changes can disrupt these conditions, leading to plant stress or death.

Experimental manipulations of absolute concentrations of micro- and macronutrients that otherwise maintain their relative concentrations should have less effect on plant growth and fitness than experimental manipulations of absolute concentrations of resources like light, water, or CO_2. However, the opposite pattern is expected when manipulating relative nutrient concentrations. Organisms that maintain strict internal homeostasis are predicted to maintain constant stoichiometric ratios of their tissues regardless of the absolute or relative availability of resources in the environment (Sterner and Elser 2002). On the other hand, tissue concentrations of nutrients in organisms with no homeostatic control—the "you are what you eat hypothesis" (Persson et al. 2010; Elser et al. 2010)—will mirror those of the environment and should be affected little, if at all, by changes in total resource availability that maintains constant stoichiometric ratios.

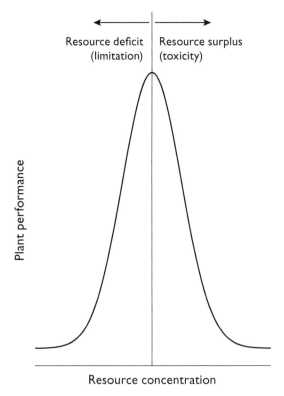

Resource deficit (limitation) | Resource surplus (toxicity)

Plant performance

Resource concentration

FIGURE B.1. Unimodal response curve of plant performance as a function of the concentration of a resource. At low levels, resources are limiting, so plant performance increases as the concentration increases. Beyond an optimal level, the resource becomes toxic at high concentrations, so plant performance declines. Although the graph is depicted with the absolute resource concentration on the x-axis, the shape of the curve for nutrient resources may depend on the relative concentration and the ratios of limiting nutrients.

Most crop plants are grown under conditions of resource excess, and additional growth is controlled more strongly by nutrient ratios (Mueller et al. 2012). Plants in more natural settings, however, may experience chronically low concentrations of both kinds of resources. Their growth will be more sensitive to absolute resource levels (Chapin et al. 1986). When concentrations become high enough, all resources start to become toxic, so a bell-shaped response curve should be apparent for any plant trait or measure of plant performance that is plotted against a sufficiently wide range of resource concentrations (figure B.1). Coordinated responses by suites of traits to absolute or relative nutrient concentrations ultimately translate into life-history transitions (growth, mortality, flowering, fruiting, seed set) that determine population growth rate and the chances for long-term persistence or local extinction (chapters 5–7).

Deterministic Stage-Based Models

Matrices of population age-, size-, or life-history stage are the foundation of demographic analysis, and can be analyzed with either discrete-time mathematics (e.g., Caswell 2006) or continuous integral projections (Ellner and Rees 2006). we favor the discrete-time, stage-based approach for its simplicity, utility, and generality.

A discrete-time matrix population model takes the form

$$n_{t+1} = An_t, \tag{C.1}$$

where n_t is a column vector with elements n_i corresponding to the number of individuals of age, size, or life-history stage i at time t and A is a square $n \times n$ Lefkovitch matrix (Lefkovitch 1965), with entries F_j in the first row corresponding to the number of offspring produced by individuals in each stage j and other entries a_i, j corresponding to the probabilities that an individual of (st)age i at time t will move ("transition to") (st)age j at time $t + 1$ (Caswell 2006):

$$A = \begin{bmatrix} a_{1,1} & F_2 & F_3 & \dots & F_n \\ a_{1,2} & a_{2,2} & a_{2,3} & \dots & a_{2,n} \\ \vdots & \vdots & \vdots & \ddots & \vdots \\ a_{1,n} & a_{n,2} & a_{n,3} & \dots & x_{n,n} \end{bmatrix} \tag{C.2}$$

In an age-structured population, all the off-diagonal entries in rows 2–n would equal 0. But in a stage-structured population, off-diagonal entries may be nonzero. Stage-based demographic models also have more interesting possibilities and dynamics—especially for perennial plants and other clonal organisms (Jackson et al. 1986)—than simpler, age-structured models. In the latter, each node represents the age of a single individual, so the only possible transitions are from one age to the next (individuals either grow one year older, or they die), and from reproductive adults to new juveniles. However, in stage-structured models, individuals in each of the stages except the recruitment stage can remain in the same stage for two or more consecutive years.

The population vector of (st)ages at time $t + 1$ (n_{t+1}) is obtained by left-multiplying the transition matrix A by the population vector at time t (n_t). Although the vector n changes at each time step, the transition matrix A is a constant. In the absence of environmental change, density dependence, or mutation, it does not change through time.

This simple model (equation C.1) is deterministic, and after only a few time steps, the population will achieve a stable (st)age distribution, with constant proportions of individuals n_i in each age, size, or stage class. This constant vector of proportions, w, is the right eigenvector of the Lefkovitch matrix A. Depending on the initial distribution of stages at $t = 0$, there can be a period of transient dynamics, but the population quickly settles into a stable (st)age distribution and the population grows exponentially.[1]

In the deterministic model (equation C.1), the eigenvalue of A is the finite population growth rate. It can be iteratively computed as $\lambda = \frac{N_{t+1}}{N_t}$, where $N_t = \Sigma n_t$ and converges to the solution of $|A - \lambda I| = 0$, where I is the identity matrix. The intrinsic rate of increase of a population, r, can be estimated as $\ln(\lambda)$. Under many circumstances, λ also equals the mean fitness of the population and each transition probability (reproduction, survivorship, stage transitions) can be interpreted as an individual component of fitness (van Tienderen 2000).

Since the introduction of age-based Leslie matrices into ecology textbooks in the 1960s and 1970s, and the publication by John Harper (1977) of *The Population Biology of Plants*, much effort has been devoted to estimating transition matrices and λ values for populations of many plant and animal species (Salguero-Gómez et al. 2015). Estimates of λ also have been used to predict probabilities of persistence or extinction for threatened species (e.g., Ehrlén and Morris 2015; Hutchings 2015).

Elasticity analysis determines the sensitivity or relative importance to λ of changes in each transition probability (van Tienderen 2000; Franco and Silvertown 2004). Given the previously defined left eigenvector w (the column vector of proportions of individuals in each (st)age class once the stable stage distribution is reached), and the right eigenvector v (vector of reproductive values of each (st)age of A), the elasticity e_{ij} of each element of A is calculated as

$$e_{ij} = \frac{a_{ij} s_{ij}}{\lambda}, \tag{C.3}$$

where the sensitivity s_{ij} of each element of the matrix equals $\frac{v_i w_i}{w \cdot v}$ (for additional analytical details, see Caswell 2006). Elasticity analysis has been used to identify particular (st)ages that have large effects on population growth rates and to prioritize these (st)ages for management interventions.

We emphasize that the assumption that A is constant in equation C.1 is rarely true. Projection models based on this assumption can give misleading forecasts because they ignore two major sources of uncertainty: uncertainty caused by changing environmental conditions and uncertainty caused by measurement error and imprecise estimates of the elements of the transition matrix. As the pace of global environmental change increases, it is important to include both environmental change and measurement error in demographic modeling (e.g., Åberg 1992; Keith et al. 2008; Ehrlén and Morris 2015).

The Basics of Species Distribution Models

D.1 FORECASTING SPECIES DISTRIBUTIONS

Species distribution models (SDMs) combine data on where a species occurs with climatic data to identify a climatic "niche" for the species. Since 2005, two data sources have driven the modeling and forecasting of how species distributions will change with climate. The first data source is species occurrence data. These data most commonly are "presence-only data" derived from georeferenced specimen records of species occurrences (Bachman et al. 2011) in museum collections (Graham et al. 2004) or ecological surveys (Lawton et al. 1998). The second data source are climatic data derived from global maps of a variety of climatic variables and conveniently available from sources such as PRISM (Prism Climate Group 2004) or WorldClim (Fick and Hijmans 2017). These global maps have been interpolated and estimated from data collected at weather stations and can be downscaled to rasters that are linked to the specimen or survey data. These global maps also are linked to climate-forecasting models to give spatially gridded forecasts (or hindcasts) of future (or past) climatic landscapes (e.g., Nogués-Bravo 2009).

These two data sources are combined in various models to estimate the probability of suitable habitat where one or more species could occur, given current or future climates. These forecasts routinely behave like a two-dimensional conveyor belt, with species ranges in a warming climate shifting toward higher latitudes and elevations or shrinking to the point of local or regional extinction (e.g., Fitzpatrick and Ellison 2018; Steinbauer et al. 2018). However, the temporal trajectory of species richness and biodiversity change may not be smooth. For example, species with similar climate niches are clustered in space and there may be abrupt shifts through time in species richness and composition with forecasted changes in temperature and precipitation (Trisos et al. 2020).

D.2 MODELING FRAMEWORKS FOR SDMS

Several statistical modeling frameworks are used for SDMs, but by far the most frequently used is MaxEnt ("Maximum Entropy"), a machine-learing algorithm that

TABLE D.1. Format for data input into a basic species distribution model (SDM).

Species	Longitude	Latitude	Covariate(s)	...
Text	Decimal-degrees	Decimal-degrees	Value(s)	...
⋮	⋮	⋮	⋮	⋮

optimizes the fit of species occurrence data to measured environmental variables (Phillips et al. 2006; Elith and Leathwick 2009). Part of the popularity of MaxEnt stems from the relatively simple data structure of its inputs: a data frame (spreadsheet; table D.1) in which each species occurrence is a row and the columns include geographical location (latitude, longitude) and environmental covariates (discrete or continuous biotic or abiotic variables measured at each location). Many datasets of species occurrences have been aggregated and are easily accessible from data portals such as GBIF[1] and iNaturalist,[2] among many others.[3]

Fitting an SDM requires knowledge not only of where a species occurs within its range but also where it does not occur. If true species absences are available for a set of geographical coordinates within a species' range, a logistic regression or other general linear model would be the best method with which to describe the relationship between environmental covariates (such as temperature, precipitation, or N deposition) and the probability of species occurrence (Peterson et al. 2011). But such absence data usually are not available unless investigators monitor entire assemblages instead of individual species (Dornelas and The BioTIME Consortium 2018).

MaxEnt works around these data gaps by imputing random, uniformly distributed pseudoabsences within a specified background area (input raster) provided by the user. The augmented dataset consisting of real presences plus pseudoabsences is then repeatedly partitioned into testing and training sets for optimizing model fit. The end result is a habitat suitability index for each location in the dataset. This index (scaled from 0.0 to 1.0) measures the probability of environmental conditions, given that a species occurs there (Phillips et al. 2006). However, ecologists routinely interpret it the other way around: the probability of species occurrence, given the presence of environmental conditions (Royle et al. 2012; Yackulic et al. 2013).

Although these two conditional probabilities are not at all equivalent, the latter interpretation (P(occurrence | environmental conditions)) is useful for comparison with the output from a demographic model. In particular, stochastic simulations of λ, the finite rate of increase, can be used to estimate the long-term probability that a population goes extinct at a given site. However, as we show in chapter 7

(§7.3.2), MaxEnt and demographic modeling generate very different forecasts for the fate of *Sarracenia* populations across the United States.

More recent and sophisticated species distribution models incorporate the presence of other species (competitors, predators, parasites, or mutualists) as potential biotic drivers of species occurrence (e.g., Pollock et al. 2014; Lany et al. 2018). However, these forecasts are complicated by the limited dispersal potential of some taxa (including *S. purpurea*; appendix A, §A.4.2) that may not be able to move fast enough to keep pace with the "velocity" of climatic change (Svenning and Skov 2004; Loarie et al. 2009). Other approaches have focused on genetic variability within and between populations (Gotelli and Stanton-Geddes 2015), which may either limit or enhance the spread of species in a warming world (Sork et al. 2010; Fitzpatrick and Keller 2015).

A Brief History and Précis of Methods for Analyzing Ecological Communities

The ideas and results we presented in chapters 9 and 10 derive and extend from more than a century of research and analytical developments about patterns of, and processes determining, species co-occurrence, food-web structure and dynamics, and succession. To a large extent, this body of "MacArthurian" ecology has largely defined the domain of academic community ecologists since the early 1960s. However, plant and animal ecologists have pursued these topics somewhat in isolation, emphasizing slightly different aspects of community assembly and organization. The methods we present here for organizing and analyzing community-ecology data reflect similarities and differences in a conceptual or intellectual [plant or animal ecologist] × [co-occurrence or food web or succession] matrix.

E.1 SPECIES CO-OCCURRENCE AND ASSEMBLY RULES

Like the OED (OED Online 2018a) and some of the earliest ecologists (e.g., Forbes 1887; Elton 1927), we begin with communities of animals. Animal community ecologists emphasized the importance of biotic interactions—especially competition and predation—as determinants of observed patterns of individual morphology; resource use by coexisting species; and local species richness, species composition, and relative abundance (Wiens 1989). Factors emphasized by plant community ecologists, such as environmental variation, land-use history, and disturbance, were not ignored by animal ecologists, but the problem of coexistence among animal species was framed as a problem of niche overlap and the ability of species to persist while competing for food and space (Cody and Diamond 1975; Chase and Leibold 2003).

In his final book, *Geographical Ecology*, MacArthur (1972) argued that processes determined by local interactions could be scaled up to predict larger-scale patterns in geographical ranges and species co-occurrences. In a memorial volume edited and written by MacArthur's colleagues after his death, Diamond (1975) extended MacArthur's proposition and introduced the idea of community "assembly rules." Although "assembly rules" *prima facie* imply temporal dynamics (analogous to the mechanisms of succession described concurrently by Connell and Slatyer 1977), Diamond (1975) ignored time in his explanation of community structure at equilibrium. Diamond (1975) also restricted his analysis to functionally similar groups of related species (e.g., fruit-eating pigeons) that were at a single trophic level (i.e., taxonoic guilds). He largely ignored other interactions, including predation, parasitism, pollination, and combinations thereof (e.g., intraguild predation; Polis 1991).

Diamond's assembly rules extrapolated local mechanisms of niche segregation to geographic patters of occurrences of birds on islands. For a geographically defined group of S species (e.g., the birds of the New Hebrides [now Vanuatu] archipelago), there are $2^S - 1$ possible combinations of species that could occur, Not all do (or were observed), and Diamond (1975) assumed that the unobserved combinations were "forbidden" by species interactions. Diamond (1975) also described several examples of "checkerboard distributions": pairs of ecologically similar congeners that never co-occurred on the same island. More generally, Diamond (1975) argued that competitive interactions and subsequent niche adjustments could account for any pattern of observed species co-occurrence.

E.1.1 The Presence–Absence Matrix

The core data structure for analyzing species co-occurrence and identifying assembly rules (sensu Diamond 1975) is a binary presence-absence matrix (McCoy and Heck 1987), in which rows are species, columns are sites or samples, and the entries are the presence (1) or absence (0) of a species in a sample (table E.1).

TABLE E.1. A hypothetical example of a binary presence–absence matrix, with rows for species, columns for sites, and entries representing the presence or absence of a species at a site.

	Site							
	S_1	S_2	S_3	S_4	S_5	S_6	S_7	$\cdots$
Species A	0	1	0	1	0	1	0	.
Species B	1	0	1	0	1	0	1	.
Species C	1	1	1	1	1	0	0	.
Species D	1	1	1	0	0	0	0	.
Species E	0	0	0	1	1	1	1	.
⋮	.	.	.	.	.	.	.	⋱

Species pairs (AB) and (DE) are two examples of perfect checkerboards, whereas the distribution of species C is nested (sensu Patterson and Atmar 1986) within that of species D.

The structure of the presence–absence matrix (table E.1) reflects important assumptions and decisions by investigators about how communities should be sampled. The matrix rows represent the set of species chosen, bounding the taxonomic and trophic limits of the community. The matrix columns represent the sites (or samples), which have an explicit spatial scale determined by the grain size and extent of the sampling. Because the entries in the presence–absence matrix consist only of ones and zeroes, the problem of estimating population sizes is avoided. However, presences or absences of species still are affected by an implicit definition of species occupancy, which usually does not allow for the possibility of detection errors (Royle and Dorazio 2008).

E.1.2 Quantifying Co-occurrence

Although dozens of highly correlated species co-occurrence metrics have been proposed (Arita 2017), the C-score (Stone and Roberts 1990) is a widely used index that has good statistical power for detecting patterns of nonrandomness in the distribution of presences and absences (Gotelli 2000). For each unique species-pair ij in the presence–absence matrix (table E.1), the C-score for that pair is calculated as

$$C = (R_i - S) \times (R_j - S), \qquad (E.1)$$

where R_i and R_j are the number of sites (e.g., islands or individual *Sarracenia* pitchers) occupied by species i and j (i.e., the row totals for species i and j), and S is the number of sites that contain both species. The C-score ranges from a minimum of 0 (all sites shared) to a maximum of $R_i R_j$ (no sites shared, in which case the two species form a perfect "checkerboard" (sensu Diamond 1975). The overall C-score for a site × species presence-absence matrix is the average of all the C-scores for all possible pairs of species. Large C-scores indicate a small number of pairwise species co-occurrences (i.e., species segregation), whereas small C-scores indicate a large number of species co-occurrences (i.e., species aggregation). Intermediate scores reflect either a random pattern of co-occurrence or a mixture of highly aggregated and highly segregated pairs within the same matrix.

E.1.3 Null Models for Co-occurrence

Diamond (1975) emphasized the operation of competition in the assembly rules for New Hebridean birds and, by implication, to most interacting taxa on any island ecosystems. Connor and Simberloff (1979) argued that a handful of carefully selected examples from a large tropical avifauna could be found that would support almost any pattern of co-occurrence, and that Diamond's assembly rules represented post hoc interpretations lacking predictive power (Gotelli 1999). The subsequent acrimonious debate precipitated the development of formal null hypotheses and associated null models to test whether observed patterns of species co-occurrence differed from those expected in the absence of species interactions (Gotelli and Graves 1996). This work extended pioneering studies by Pielou (Pielou and Pielou 1968; Pielou 1972), who not only had described patterns of co-occurrence among spiders, but also provided detailed statistical tests of an explicit null hypothesis based on independent species occurrences. After nearly 20 years of vitriolic exchanges following the publication of Diamond (1975), a meta-analysis of 96 published presence–absence matrices produced results that were consistent with Diamond's (1975) assembly rules (Gotelli and McCabe 2002). Independent of spatial grain, extent, and latitude, the published matrices as a group exhibited more checkerboard species pairs, fewer species combinations, and more pairwise species segregation than expected by chance. However, these patterns were much stronger for bats, mammals, birds, ants, and plants and not different from random for invertebrates, fish, reptiles, and amphibians.

With a null-model analysis, we ask what values of the C-score (or any community metric) would be expected if the species occurrences among sites are random and independent of one another. If species on islands or in patches are nonrandomly distributed, the contrasting null model should include random and independent colonization effects as in the equilibrium model for island biogeography (MacArthur and Wilson 1963) or later neutral models (Caswell 1976; Hubbell 2001).

Because the parameters for those models are almost impossible to estimate from static presence–absence matrices (Gotelli and McGill 2006, but see Ricklefs 2003), ecologists turned to randomization tests that used information from the binary presence–absence matrix itself (table E.1) to construct null distributions (Gotelli and Ulrich 2012). But the devil is in the details. There are many different ways to randomize a presence–absence matrix to mimic the effects of random colonization and extinction (Strona et al. 2018), but three algorithms—fixed–equiprobable, fixed–fixed, and fixed–proportional (Gotelli 2000)—have stood the test of time.

Fixed–Equiprobable   In this algorithm, the occurrences of each species are independently reshuffled among the sites. This model fixes the row totals of each null matrix to match the row totals of the original (observed) matrix. Fixing the row totals preserves the relative commonness and rarity of each species in the null-model simulations. However, the column totals (number of taxa per site) are allowed to vary randomly and freely. Thus, the sites are treated as equiprobable and observed differences among sites in total species richness (i.e., column totals) are not preserved in the simulations. This model is equivalent to a simple "balls-in-urns" model and corresponds to a 2×2 contingency analysis (Arita 2016).

This model has good statistical properties (Gotelli 2000) unless there are marked differences in the suitability of each site for species occupancy. Such differences in suitability might be caused by variation in, for example, the areas, resources, habitats, abiotic factors, or history of the sites.

Fixed–Fixed   In this second algorithm, both the row totals and the column totals of the original matrix are preserved. Thus, the simulated matrices contain the same number of occurrences per species (row totals) and the same number of species per site (column totals) as the original matrix. Nonrandom co-occurrence patterns are measured above and beyond the baseline patterns that are determined by these marginal constraints. Because so many species diversity and species co-occurrence metrics are sensitive to species richness and relative abundance (Ulrich et al. 2018), this algorithm has good type I error properties and does not lead to incorrect rejection of the null hypothesis for random matrices with heterogeneous row and column sums (Gotelli 2000; Strona et al. 2018).

However, the good type I error properties of the fixed–fixed algorithm are offset by its greater likelihood of type II errors: incorrectly accepting a null hypothesis that is false. type II errors can be a problem in very small matrices. When there are few rows and columns, there may be only a limited number of randomizations possible that preserve both row and column totals. Similarly, if the row or column totals are either very small (a minimum of 1) or very large (a maximum of the number of rows or columns), there also may be a limited number of possible matrix rearrangements.

Moreover, fixing both row and column totals imposes a kind of zero-sum constraint on the randomizations. As a consequence, some first-generation simulation algorithms produced biased matrices in which occurrences of each species were not strictly independent of the occurrence of other species (cf. Zaman and Simberloff 2002). Strona et al. (2014) addressed this problem with a "curve-ball algorithm" that efficiently shuffled elements for pairs of rows and generated an unbiased set of random matrices with the same row and column totals.

Fixed–Proportional   The third algorithm is similar to the first (fixed–equiprobable), but the assignment probabilities for the sites are based on an independent measure of site suitability rather than just an assumption that all sites are equiprobable. This is analogous to a "balls-in-urns" model in which the urns are of unequal sizes.

C-scores and other metrics used with these three different null models can give contradictory results (e.g., Gilpin and Diamond 1982). Tests of different null models with artificial datasets containing specified levels of structure and noise also have revealed that different null models have different risks of type I or type II statistical errors (e.g., Gotelli 2000; Hardy 2008; Harris 2016). Now, nearly a half-century after Diamond (1975) introduced the concept of assembly rules, ecologists have a continuous landscape of possible null models from which to choose one or more to test the rules' significance (Gotelli and Ulrich 2012; Strona et al. 2018).

E.1.4 Pairwise Analyses

The structure of entire assemblages represented by a presence–absence matrix ranges along a continuum from strongly segregated (a perfect checkerboard) to strongly aggregated (species completely nested, sensu Patterson and Atmar 1986) (Leibold and Mikkelson 2002). However, analyses of co-occurrence that rely on a single metric such as the C-score may distort or oversimplify complex patterns that reflect interactions within and among subsets of species (Grant and Abbott 1980). In fact, the same matrix can be shown to be both significantly nested and significantly aggregated (Ulrich and Gotelli 2007b).

These considerations motivated the analysis of individual species pairs within a matrix using the same null-model algorithms that had been applied to the entire matrix (Sanderson 2000; Sfenthourakis et al. 2006). However, these pairwise analyses calculate an index of species co-occurrence for each species pair, resulting in $(S \times (S - 1))/2$ significance tests rather than a single matrix-wide test. This multitude of tests can produce many potential false positives, a problem not only for ecology but also for analyses of gene expression and microarrays (Kammenga et al. 2007). Efron (2005) provides tools for reducing the number of false positives and preserving an overall family-wide type I error rate of 0.05.

Early critics of null models also worried about a "dilution effect": a matrix could appear random because the important interactions between pairs of competitors were being diluted by all of the random associations between pairs that were not sharing resources or interacting (Grant and Abbott 1980). To address this, Gotelli and Ulrich (2010) adapted an empirical Bayes approach of binning each species pair and then screening nonrandom pairs according to the frequency of such pairs generated by the null model. Thus, each species pair in the matrix can be classified as aggregated, random, or segregated. The results of such pairwise analyses were unexpected: more than 95% of the species pairs in most matrices were randomly associated (Gotelli and Ulrich 2010). Of the small number of pairs that passed the Bayes' criterion test for nonrandomness, most were negative rather than positive associations (although the reverse seemed to be true for fossil assemblages before the mid-Holocene; Lyons et al. 2016). Thus, most matrices reflect a "concentration effect": the community level nonrandomness of most presence–absence matrices often reflects strong nonrandom associations between only a handful of species pairs.

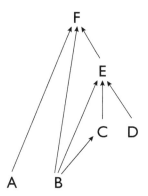

FIGURE E.1. A food web with six species and four trophic levels. Arrows go from prey to predators.

E.2 FOOD WEBS AND OTHER NETWORKS

Co-occurrence analyses and assertions of assembly rules usually have been re-stricted to taxonomic or functional groups of species at the same trophic level. A much richer array of direct and indirect interactions are possible for analyses of entire food webs, which consist of unrelated species that acquire resources in many different ways.

E.2.1 Food-Web Matrices

Food webs usually are illustrated as directed networks (a.k.a. directed acyclic graphs, or DAGs), in which individual species (or functionally similar groups of species such as "ants") are depicted as nodes. Interactions between species in food webs are trophic interactions ("who-eats-whom" and at what rate; Ulanowicz 2004), which are depicted as edges (figure E.1). A network of nodes and edges can be summarized in matrix form: rows and columns are individual nodes, and entries represent the presence (1) or absence (0) of an edge between them (table E.2). This food-web matrix (or "adjacency matrix" in Ulanowicz 2004) is a special and much simplified case of two different types of matrices used respectively by ecologists studying populations or communities (Levins' 1968 community matrix) and those studying ecosystem dynamics (Ulanowicz's 2004 matrix of dietary proportions and its relatives).

The Community Matrix For a set of S species, each element α_{ij} of the $S \times S$ community matrix is the per capita effect of species i on the population growth rate of species j; the diagonal elements are the per capita effect of each species on itself (table E.3).

TABLE E.2. A matrix representing the simple food web shown in figure E.1.

	Species					
	A	B	C	D	E	F
Species A	1	0	0	0	0	1
Species B	·	1	1	0	1	1
Species C	·	·	1	0	1	0
Species D	·	·	·	1	1	0
Species E	·	·	·	·	0	1
Species F	·	·	·	·	·	1

Rows and columns represent species (nodes), and 1s represent interactions (edges). By convention, interactions flow from rows ("prey") to columns ("predators"). Non-zero entries along the diagonal imply some forms of intraspecific control on population dynamics (e.g., density-dependence or cannibalism).

TABLE E.3. Levins' (1968) community matrix.

	Species		
	A	B	C
Species A	α_{AA}	α_{BA}	α_{CA}
Species B	α_{AB}	α_{BB}	α_{CB}
Species C	α_{AC}	α_{BC}	α_{CC}

Each entry is the interaction coefficient α_{ij}, which measures the per capita effect of species i on the population growth rate of species j.

In the community matrix, the population dynamics of all species are directly and indirectly linked through a set of simultaneous first-order differential equations. These equations generalize the two-species Lotka–Volterra competition model to multiple species (Levins 1968; Wootton 1998). Positive, negative, and zero values for the α's in the community matrix respectively imply mutalistic, competitive, or neutral inter- and intraspecific interactions. Indirect effects are propagated through the entire community because the equations for all species are coupled. The growth equations can be solved to yield equilibrium population sizes, and the first eigenvalue of the community matrix measures local (Lyupanov) stability: the rate of return to equilibrium after a small perturbation (Caswell 2006).

The Matrix of Dietary Proportions   Scientists who study flows of energy and nutrients through ecosystems replace species in food webs with compartments representing functional groups, such as plants (or primary producers), detritus, detritivores, herbivores, and carnivores. Within-compartment dynamics (i.e., the diagonal elements of table E.3) are ignored and effects of one compartment on another are measured as trophic exchanges of energy (e.g., $kcal \cdot m^{-2} \cdot yr^{-1}$) or nutrients (e.g., $kg\ N \cdot ha^{-1} \cdot yr^{-1}$). To work with these trophic exchanges, the upper off-diagonal elements α_{ij} of the community matrix (table E.3) are replaced by elements g_{ij} of a square matrix of dietary proportions G (Ulanowicz 2004). Each element g_{ij} of G is the percent that each compartment i contributes (outputs) to the total diet (input) of the compartment j.

The individual elements of G are computed as

$$g_{ij} = \frac{T_{ij}}{T_{\cdot j}X_j},$$ (E.2)

where T_{ij} is the rate of transfer of energy from compartment or taxon i (prey) to another compartment j (e.g., its predator), and X_j is the rate of any exogenous inputs to taxon j. The overall matrix of dietary proportions G is thus an $S \times S$ matrix with upper diagonal elements g_{ij}, diagonal elements = $\{0\}$, and lower diagonal elements normally = $\{0\}$ (unless prey become predators at, for example, different life-history stages).

The G matrix has two interesting properties. First, the elements of G^m represent the trophic contributions of every path of length m through the web. Second, the infinite series $I + G + G^2 + G^3 + \dots$, where $bmI = G^0$ = the identity matrix, converges to the "Leontief structure matrix" $S = [I - G]^{-1}$ (Leontief 1951), whose ijth entry is the fraction of the total input to species or compartment j that comes from species or compartment i over all pathways of all lengths.

The S matrix also can be used to determine intermediate transfers between compartments. Szyrmer and Ulanowicz (1987) transformed S into another matrix D (the "total dependency matrix") whose elements d_{ij} are interpreted as the fraction of the total inputs (diet) of compartment (predator) j that passed through compartment (prey item) i before it was consumed by j:

$$d_{ij} = (s_{ij} - \delta_{ij}) \left(\frac{\sum_{k=1}^{n+2} T_{ik}}{s_{ii} \sum_{m=0}^{n} T_{mj}} \right).$$ (E.3)

Here, s_{ij} are elements of S, δ_{ij} are elements of the identity matrix, T_{0j} are the external inputs to compartment j, $T_{i,(n+1)}$ are usable exports from compartment i, and $T_{i,(n+2)}$ are dissipative losses (e.g., respiration) from compartment i. Another way of interpreting D is to think of each of its columns as the "indirect" input (diet) of compartment (predator) j (Ulanowicz 2004). Other normalizations of the T matrix (whose elements T_{ij} are all the flows between compartments), such as row-wise normalizations instead of the column-wise normalizations in D, yield additional information about energy transfer through different pathways (Ulanowicz 2004).

E.2.2 Constructing Food Webs

It is very difficult to estimate the coefficients of a community matrix (table E.3), reconstruct the associated interaction network (figure E.1, table E.2), or measure energy flows g_{ij} of G. Comprehensive pairwise experiments that measure per capita density effects of species on one another are used to fill the community matrix (Paine 1992; Berlow et al. 1999). Long-term observations and measurements of ecosystem processes can estimate g_{ij}, but it is rare that all compartments are measured simultaneously (e.g., Tilly 1968; Baird and Ulanowicz 1989).

More typically, food-web data are compiled from biological knowledge of predator–prey relationships, aggregation of multiple studies, or inferred from quantitative surveys of species associations. The latter may include data on species co-occurrences (table E.1; Harris 2016) and covariation in abundance. When tested against prespecified food-web structures, these methods do a better job of recovering the food-web structure than piecing together the results of a pairwise co-occurrence analysis, which simply emphasizes the statistical pattern of pairwise aggregation, segregation, or randomness. However, Harris's (2016) method does not seem to control for sampling effects, especially those that introduce heterogeneity into row and column sums of the matrix.

E.2.3 Metrics and Properties of Food Webs and Other Networks

Once food webs have been assembled, their structure or trophic dynamics can be quantified. Standard measures of food-web structure and dynamics include metrics that are general properties of networks and others that are more specific to trophic webs. Table E.4 summarizes the representative set of whole-web metrics that we used in our analysis of the inquiline food webs of *S. purpurea*. A more comprehensive list of network metrics—many of which are correlated with or derived from one another—is given by Lau et al. (2017) and implemented in the enaR package (Borrett and Lau 2014).

Early studies of food webs and network properties figured prominently in the evaluation of diversity–stability relationships (Elton 1958). MacArthur (1955) argued that stability of a more highly connected network would be higher because there would be more potential pathways for energy flow, as in complex small-world networks (Watts and Strogatz 1998). But initial explorations of this hypothesis led to a different result. When the interaction coefficients in the community matrix (table E.3) were drawn from a random uniform $[-1, 1]$ distribution, Lyupanov stability decreased as more links were introduced (Gardner and Ashby 1970). May (1972) confirmed that average interaction strength would have to decrease as the number of species and connectance of the web increased.

TABLE E.4. Metrics of whole network and food-web structure used commonly by ecologists.

Metric	Meaning	Applies to
N or S	Number of nodes (species)	Networks
L	Number of edges (interactions)	Networks
$LD = L/S$	Average link density	Networks
$LD_E = e^{\frac{H_c}{2}}$	Effective link density[a]	Trophic webs
$C = L/[S(S-1)/2]$	Connectance (a.k.a. density)[b]	Networks
TD	Trophic depth (# of trophic levels)	Trophic webs
$TD_E = e^{AMI}$	Effective trophic depth[c]	Trophic webs
$A = \sum_{i,j} T_{ij}\log\left(\dfrac{T_{ij}T_{..}}{T_{i.}T_{.j}}\right)$	Network ascendency	Trophic webs

After Ulanowicz (2004) and Lau et al. (2017).

[a] $H_c = -k\Sigma_{i,j}\frac{T_{ij}}{T_{..}}\log\left(\frac{T_{ij}^2}{T_{i.}T_{.j}}\right)$ is the residual diversity of network flow (sensu Rutledge et al. 1976, after MacArthur 1955).

[b] If connectance includes cannibalism and mutual predation, then $C = L/S^2$ (Martinez 1992).

[c] $AMI = k\Sigma_{i,j}\frac{T_{ij}}{T_{..}}\log\left(\frac{T_{ij}T_{..}}{T_{i.}T_{.j}}\right)$ is the average mutual information (sensu Ulanowicz 2004, after Rutledge et al. 1976).

However, some of the simulated networks had negative population sizes at equilibrium, which is not possible in nature (Roberts 1974). Others had trophic chains that were so long that they defied principles of limited energy transfer from one trophic level to the next (Lindeman 1942). Still others had intransitive network loops (e.g., A affects B, B affects C, C affects A; Gilpin 1975), which contradicted the view that competitive interactions formed transitive linear hierarchies (Keddy and Shipley 1989; but see Buss and Jackson 1979). However, as food web compartments ("decomposers," "microbes," "plants," "omnivores," etc.) have become better resolved, long trophic chains, widespread omnivory, cannibalism, intraguild predation, and other complex interactions have emerged as the norm rather than the exception (e.g., Polis 1991; Banasek-Richter et al. 2009).

Major advances in network theory and computational analysis over the past 20 years have emphasized local rather than global stability (Grilli et al. 2017) and the importance of clustered subunits embedded in larger interaction networks (Watts and Strogatz 1998). Ecological guilds of species sharing a common focal resource may represent examples of small-world networks, in which there are tight clusters of interacting nodes with some links connecting them to other such clusters (Krause et al. 2003). Compared to a randomly connected network (i.e., an Erdös–Rényi network Erdös and Rényi 1959), small-world networks generate a fat-tailed distribution of linkage distances, which may follow an approximate power distribution with a scale-free exponent of between 2 and 3 (May 2006). However, there is still uncertainty over whether network metrics are scale invariant (Martinez 1992; Keller 2005), and how much network patterns are influenced by incomplete sampling (Blüthgen 2010).

Recent analyses of more detailed food-web models suggest that interacting subunits may impart stability to large ecological networks (Rooney and McCann 2012; Allesina and Tang 2012). Weak interactions (McCann et al. 1998), and interactions with rare species (Yoshimura et al. 2006). may stabilize network structure. The stability of a large network may be sensitive to the loss of particular interactions (Allesina and Pascual 2008; Allesina and Tang 2012) or highly linked nodes (O'Gorman and Emmerson 2009; Ives and Cardinale 2004) rather than the average connectance of the entire network (Martinez 1992; Olesen et al. 2011).

E.3 COMMUNITY SUCCESSION

Co-occurrence analysis and food-web metrics consider communities to be in equilibrium. Although explicit analysis of spatial variation in animal communities and food webs is common (Menge 1995; Cottenie 2005), temporal changes in their structure have been considered only infrequently (Drake et al. 1993; Morton et al. 1996; López et al. 2018; Hutchinson et al. 2019; Ohlmann et al. 2019). Plant ecologists, however, have been studying temporal dynamics under the rubric of community succession since the late nineteenth century (Cowles 1899, 1911; Clements 1916).

Succession is the systematic change in the composition and abundance of species through time after creation or exposure of new land or soil ("primary succession" after, e.g., volcanism or glacial retreat; Del Moral and Bliss 1993; Fastie 1995) or following a disturbance that has removed most or all of the species and biomass in a patch ("secondary succession" after, e.g., creation of a tree-fall gap; Pickett and White 1985; Poulson and Platt 1996, figure E.2). The same James Hervey who coined the term "community" in 1747—lending the epigraph to chapter 8—also wrote in the same volume about succession: "Another remarkable Circumstance, recommending the flowery Creation [the annual cycle of flowering herbs], is their *regular Succession*" (Hervey 1746: 57; italics in the original).

Two empirical observations of terrestrial vegetation motivated definitions and study of succession. The first observation was that, following a disturbance, a distinctive set of species ("pioneers") would establish (Cowles 1911). These species often had morphological and physiological adaptations to harsh abiotic conditions (often high temperatures and light, low moisture and nutrients). The second observation was that, with enough time, replicate patches in similar environments

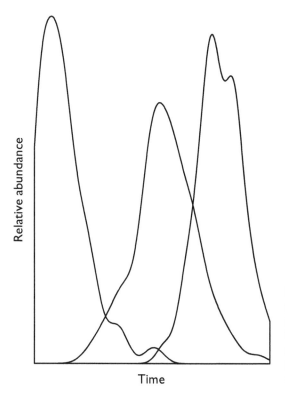

Figure E.2. The temporal progression of the entry of species into an assemblage, followed by changes in their abundance and their eventual disappearance, form the core data for studies of ecological succession.

Relative abundance

Time

that were disturbed by different forces (fires, logging, storms) would converge in species composition (Warming 1895, cited in Cowles 1911).

These observations led Clements (1916) to propose what Connell and Slatyer (1977) called the "facilitation model" of succession: the first set of species to colonize a barren or disturbed site would alter the physical environment in a way that changed the abiotic conditions and facilitated the arrival and establishment of the next set of species. This second set would replace the first and facilitate the establishment of the next. The successive replacement of species sets ("seres") would continue until a final "climax" community was reached. This climax community was originally assumed to be stable and self-replacing (and in Clements' view, analogous to a "superorganism"), with high species richness and biomass. At any stage in the sequence, however, a fresh disturbance could reset the system, followed by an orderly progression toward the climax community.

Gleason (1926) rejected Clements' view that plant succession leading to a climax community was analogous to the development of an organism. Instead, Gleason (1926) proposed that plant communities were loose associations made up of species that reflected their individual responses to the environment and one another. Although the Gleasonian view has more empirical support (Tansley 1935; Whittaker 1935; Connell and Slatyer 1977), plant community ecologists continue to identify plant community types based on a dominant species or genus (e.g., "beech-maple forest") or a dominant growth form (e.g., "grassland") (Faber-Langendoen et al. 2018). At larger spatial scales, geographers use community types to describe and map entire biomes or life zones (Whittaker 1962; Holdridge 1947; Olson et al. 2001).

The organization of species into groups that represent communities does, however, provide a useful framework that recognizes the potential importance of species interactions to generate higher-level patterns of organization (Connell and Slatyer 1977; Farnsworth et al. 2017). Note also that as the spatial grain of analysis decreases, models based on community types must collapse back to models of individual species. At a small enough spatial scale, a patch can support only a single species (or even a single individual), so the different "stages" in a community model could represent different individual species (Horn 1975; Diamond et al. 2016).

E.3.1 A Transition Matrix for Succession

The core data structure for studying and modeling succession is a species × time matrix (table E.5). In this matrix, each row is a species, and each column is a consecutive time point since the initial disturbance. The entries in the matrix represent the abundance (numbers of individuals, stems, or percent cover) of each species at each time.

Although all models of succession begin with the data structure in table E.5 (reflecting the progression of species shown in figure E.2), individual species may be grouped into sets of stages or community types. Ideally, each stage would be

TABLE E.5. Example of raw successional data.

	\multicolumn{8}{c}{Time Step}
	$\emptyset$	1	2	3	4	5	6	$\cdots$
Species A	0	10	22	44	21	11	5	·
Species B	0	28	35	5	0	0	0	·
Species C	0	0	5	50	166	156	160	·
Species D	0	0	0	0	12	37	38	·
Species E	0	0	0	0	0	1	3	·
⋮	·	·	·	·	·	·	·	⋱

The entries represent the abundance of different species at each time step, after an initial disturbance that initializes a patch to an "empty" state ($\emptyset$).

complete and mutually exclusive (and would include an initial "disturbed" state) and each species would be uniquely assigned to a single stage (analogous to the "relay floristics" model of succession proposed by Egler [1954] and reviewed by Drury and Nisbet [1973]). In practice, the classification is always messier: not all successional stages are represented in each location, each species cannot be assigned easily to a single stage, and stages rarely follow a perfectly ordered sequence in replicate patches monitored through time (Johnson and Miyanishi 2008).

E.3.2 Collecting Successional Data

Temporal trajectories of change in species (table E.5, figure E.2) and communities (table E.6) following an initial disturbance are the data underlying studies of succession, but they are difficult to collect, especially with proper replication. Although there are a few examples of long-term monitoring of the accumulation of vegetation and associated animal species following volcanic eruptions (Whittaker et al. 1989; Del Moral and Bliss 1993), many studies of temporal trajectories have not been long enough to exhibit community change (especially for terrestrial vegetation). Valid replication may not be possible because of changes in land use and climate, and many monitoring studies do not begin from the point of a well-defined disturbance or an "empty" patch (Dornelas et al. 2013). Instead, ecologists have relied on chronosequences: concurrent snapshot surveys of patches at presumed different stages of succession. Chronosequence data assume that the time since disturbance can be accurately dated for each patch, and that patches differ only in the time since disturbance. It is the second assumption that is problematic. Chronosequences may not yield the same patterns as replicate observations through time of a single patch (Johnson and Miyanishi 2008).

Empirical studies of succession in real time are possible for shorter-lived organisms. For example, Sutherland (1974) deployed small panels to study ecological succession of marine algae and fouling invertebrates over weeks and months. In

TABLE E.6. Organization of species temporal occurrences into a successional sequence of stages.

					Time Step				
	0	1	2	3	4	5	6	7	...
Stage	$\emptyset$	α	β	β	γ	γ	γ	δ	...

The community composition at each time step is assigned to a single stage in the successional model. The stages are mutually exclusive and represent all possible configurations.

contrast to the pattern of facilitation often seen in terrestrial vegetation, once an initial biofilm was established (Dang and Lovell 2016), the panels were quickly colonized, often by near monocultures of barnacles, algae, ascidians, bryozoans, sponges, or hydroids. These groups did not undergo an orderly sequence of species replacements (as correctly predicted for plants by Gleason 1926). Instead, the first arrivals persisted from their initial "land grab" and inhibited the establishment of other species until a disturbance reset the system, which might then be recolonized by a different assemblage (Law and Morton 1993). This pattern conformed to the predictions of the initial floristics model of succession proposed by Egler (1954) or the inhibition model of succession of Connell and Slatyer (1977).

E.3.3 Modeling Succession with Markov Models

Horn (1975) proposed using a Markov model to examine the dynamics of community succession. Starting with the set of n community stages in table E.6, Horn constructed an $n \times n$ matrix of the transition probabilities between all possible states at a single discrete time step (table E.7). In this successional transition matrix, the columns are the state of a patch at time t and the rows are the state at time $t + 1$. The diagonals of the matrix represent the probability of stasis (no change in patch state), and the other entries represent changes between any pair of states (which need not be symmetric). Unless the number of patches in the system is changing through time, the columns of this matrix sum to 1.0, and transitions that cannot occur directly in a single time step have a value of zero.

The different models of succession—facilitation, tolerance, and inhibition—have different numbers and placement of nonzero entries in the transition matrix. In the tolerance model, all transitions are possible in a single time step, and there

TABLE E.7. Transition matrix for a Markov model of succession.

Stage at time $(t + 1)$	Stage at time (t)				
	$\emptyset$	α	β	γ	δ
$\emptyset$	$p(\emptyset \to \emptyset)$	$p(\alpha \to \emptyset)$	$p(\beta \to \emptyset)$	$p(\gamma \to \emptyset)$	$p(\delta \to \emptyset)$
α	$p(\emptyset \to \alpha)$	$p(\alpha \to \alpha)$	$p(\beta \to \alpha)$	$p(\gamma \to \alpha)$	$p(\delta \to \alpha)$
β	$p(\emptyset \to \beta)$	$p(\alpha \to \beta)$	$p(\beta \to \beta)$	$p(\gamma \to \beta)$	$p(\delta \to \beta)$
γ	$p(\emptyset \to \gamma)$	$p(\alpha \to \gamma)$	$p(\beta \to \gamma)$	$p(\gamma \to \gamma)$	$p(\delta \to \gamma)$
δ	$p(\emptyset \to \delta)$	$p(\alpha \to \delta)$	$p(\beta \to \delta)$	$p(\gamma \to \delta)$	$p(\delta \to \delta)$

The entries represent the probability of transition from one community type to the next in a single discrete time step. In Horn's (1975) original formulation, the patches were so small that each "stage" was actually an individual species.

are no zeroes in the matrix (table E.7). The transition matrix for the facilitation model would have nonzero transition probabilities only along the diagonal (no change from one time step to the next), the first lower off-diagonal (from one stage to the next), and the top row (any stage removed following disturbance). Finally, the transition matrix for the inhibition model would have nonzero transition probabilities only along the diagonal (no change from one time step to the next), across the top row (resetting the system after disturbance), or in the first column (any stage can occur after disturbance).

As long as none of the states have an absorbing transition ($p = 1.0$), the Markov model of succession predicts a single equilibrium state in which each stage is represented in a constant relative proportion. This result is directly analogous to what we saw for demographic transition models (chapter 5), except now we are dealing with successional stages (assemblages of species) rather than life-history stages. If the Markov model is applied to data collected at a very small spatial scale, such as individual trees (Horn 1975) or ant colonies that occupy nest boxes (Diamond et al. 2016), then the stages can represent individual species.

The Markov formulation is a heuristic model because it does not explicitly incorporate processes such as inhibition or facilitation, which rather are manifest indirectly in the values of the transitions. Further, the equilibrium predictions of the Markov formulation of different successional models do not always generate unique or qualitatively different predictions. Still, the Markov model captures the essence of probabilistic successional change. It can be used with experimental and observational data and extended to incorporate density-dependence (McAuliffe 1988) and nonstationary environmental change (Tanner et al. 1996).

On Tipping Points and Regime Shifts

Tipping points are the cusp between one set of conditions ("regimes") and another; when a tipping point is passed, change is rapid, uncontrolled, and often irreversible. Passing a tipping point is like crossing a threshold from one room to another and having the door locked behind you; the state of the system after the tipping point is very different from the state of the system before the tipping point, and it is very difficult to go back.

F.1 TIPPING POINTS DEFINED

Tipping points originally were defined as "tip points" in the late 1950s to describe white flight from urban areas as African Americans moved into city centers (Grodzins 1957). Subsequently, a tipping point was also interpreted as a change in the rate of white flight from desegregated public schools to segregated private schools, not simply the process itself (Clotfelter 1976). In either case, however, the implication was that the system switched from a more preferred set of conditions to a less preferred one that, in either case, was culturally defined or conditioned.

At the turn of the twenty-first century, Gladwell (2000) redefined tipping points in his eponymous book. In doing so, he changed the way we think about rapidly emerging social phenomena, such as the Dutch tulip frenzy, the housing bubble, and the reemergence of a market for Hush Puppies (suede shoes, not puffs of fried corn meal). Although some have argued that the publication of *The Tipping Point* itself was a tipping point in how we think and talk about tipping points (Scheffer 2009; van Nes et al. 2016), Gladwell's prime analogy of a tipping point was an epidemic of a disease. Unfortunately, this analogy obscures the true meaning of tipping points, leading to much confusion in how the general public and ecologists think about responses to rapid environmental change and real tipping points.

The analogy is flawed because human disease epidemics require an initial infection of a single individual, a pool of nearby people susceptible to the same infection, and a mechanism for transmission from infected to susceptible individuals. But epidemics do not last forever. As with other classic examples of negative feedback loops, resistance evolves to new diseases and epidemics peak and burn out. Similarly, all of Gladwell's (2000) tipping point examples—the sudden emergence of fax machines and Airwalk sneakers, the resurgence of sales of Hush Puppies, the rising and then falling crime rate in many cities, and sudden epidemics of suicides—are of explosions of interest in transient phenomena followed by a return to the status quo ante and a search for the next new fad. But they are not tipping points.

F.2 TIPPING POINTS IMPLY REGIME SHIFTS

Formal, mathematical treatments of tipping points focus on rapid, seemingly permanent changes, not transient phenomena (Scheffer 2009; Petraitis 2013); in popular parlance, a tipping point presages a change in regime. Ongoing climatic change often is used as an exemplar of a tipping point in the natural world. Some researchers have asserted that we passed a tipping point in the 1980s, when human industry caused the atmospheric $[CO_2]$ to exceed 350 ppm—the so-called "safe" level of $[CO_2]$ in the atmosphere (Rockström et al. 2009; Barnosky et al. 2012, 2016). But atmospheric $[CO_2]$ has not been constant over Earth's 4.5-billion-year existence or even during the $\approx 350,000$-year history of *Homo sapiens*. Certainly the rate of change of atmospheric $[CO_2]$ has accelerated greatly since the mid-1800s. But what turns a certain level of atmospheric $[CO_2]$—280 ppm at the onset of the Industrial Revolution, 316 ppm in 1958, when Charles David Keeling started measurung $[CO_2]$ atop Mauna Loa,[1] 350 ppm in 1986, or 400 ppm in 2013—into a "tipping point"? The implication is that when a tipping point is crossed, we have shifted from one climatic regime into another.

F.3 REGIMES IMPLY ALTERNATIVE STATES

For ecologists, different regimes also are examples of "alternative stable states," and the search for them is a long-standing focus of community and ecosystem ecologists (Petraitis 2013). In a minimal system of two interacting species, more than one alternative stable state can arise in even the simplest models. For example, with certain parameter settings, the Lotka–Volterra model of two competing species yields an unstable equilibrium in which one species or the other wins, but not both. Modifying the density-dependent functions for the classic model of single-species logistic population growth to include Allee effects (inverse density-dependence at low densities) results in two equilibrium population sizes instead of a single carrying capacity (K). The larger equilibrium point is stable, but the smaller one is an unstable minimum threshold, below which the population will decline to extinction (Stephens et al. 1999).

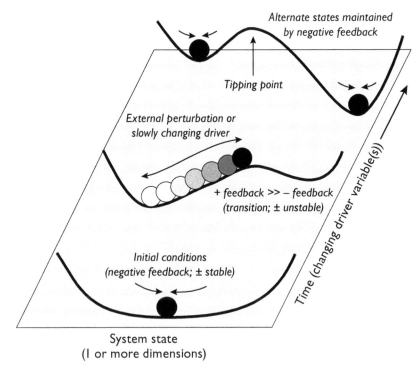

FIGURE F.1. Alternative stable states on an *n*-dimensional landscape. Initially, the system (represented as a ball) is maintained in its state by negative feedbacks operating in a particular environment. As the environment changes or the system is perturbed and the strength of positive feedbacks exceeds negative feedbacks, the system can move into a new state. The environmental landscape may change further so that negative feedbacks once again dominate and entrain the system in its new state.

A commonly used and powerful physical analogy of alternative stable states is an *n*- (usually three) dimensional landscape in which each state variable represents a different dimension. In community ecology, for example, the dimensions might represent the abundances of different species (figure F.1). At time *t*, the system can be described by a point—or better, a tiny marble—located in this coordinate space. The shape of the surface of the landscape is determined by a set of equations and their parameters. From any starting location in the space, the equations (discrete or continuous) describe where it will be at time *t*+1. The trajectory of the marble is determined by the shape of the landscape, by external perturbations that can push the system to new locations, and by the values of the state variables themselves. In this physical analogy, the alternative stable states represent valleys or low points, separated by ridges or higher points of instability.

F.4 IDENTIFYING TIPPING POINTS AND REGIME SHIFTS

Tipping points are precarious and threaten our sense of well-being. But which of the current regime or the "Golden Age" of the recent past is Dr. Pangloss's best of all possible worlds? What about the new regime we enter after we pass a tipping point? In the postcolonial world of the 1950s and 1960s, regime change—the use of military force to replace one governing administration or "hostile" foreign government by another—was usually cast in a negative light. Regime changes in which newly emerging nations shifted their alignments toward the Soviet Union or other nonaligned states were perceived as antithetical to the ambitions of the United States and NATO. The superpowers of the day did everything they could to prop up friendly regimes. In contrast, in the post-"9/11" world, regime change became desirable in some circles, especially when "rogue states" crossed "red lines." But in the context of global climatic change, regime change—following an increase of [CO_2] above 400-ppm—once again seems like a bad idea.

Academics tend to use the more neutral term "regime shift" to remove the value judgment inherent in regime change, but they still tend to view regime shifts as undesirable. As a consequence, the United States and the European Union have invested a relatively large amount of research funds on identifying early-warning indicators of impending regime shifts in global climate, economies, and societies, and developing strategies to avert them. Statisticians and modelers have developed sophisticated techniques to identify precursors to regime shifts in long time series of data (e.g., Zeileis et al. 2003; Rodionov 2005a,b; Dakos et al. 2008; Litzow et al. 2008; Boettiger et al. 2013), and the mathematical validity of these indicators has been demonstrated analytically (Sornette 2006; Scheffer 2009; Dakos et al. 2008; Scheffer et al. 2009). In parallel, ecologists have led the way in developing mechanistic models of regime shifts and indicators of them using analytical and simulation models (e.g., Carpenter and Brock 2006; Scheffer et al. 2009; Brock and Carpenter 2010; Lau et al. 2018; Ratajczak et al. 2018). However, even basic terms such as "stable" and "nonlinear" mean different things to empiricists and modelers when they link data to analytical models (Grimm and Wissell 1997; Donohue et al. 2016). The vocabulary of popular metaphors such as tipping points and regime shifts has muddied the waters even further.

F.4.1 Indicators of Regime Shifts

Following earlier work (Carpenter and Brock 2006; Dakos et al. 2008; Scheffer et al. 2009; Bestelmeyer et al. 2011; Petraitis 2013; Ratajczak et al. 2018), we suggest four empirical patterns that should be established to provide strong support for the existence of tipping points and the regimes on either side of them:

(1) multimodality in the frequency distributions of response variables; (2) rapid, nonlinear change from one definable state to another; (3) critical slowing down preceding a shift from one state to another; and (4) hysteresis in the response variable following a return of the driver to its value seen in the earlier state(s). None of these patterns by themselves is definitive, and any one of them can be generated by a variety of mechanisms that individually do not imply different, mathematically stable basins of attraction (figure F.1). Nevertheless, this list can be a worthwhile starting point for thinking about how nature is organized.

We illustrate these four empirical patterns with reference to an idealized aquatic ecosystem that can occur in one of two states an oligotrophic, clear-water "control" state, and a eutrophic, turbid-water "enriched" state. This simplified system represents a canonical example of alternative states that has been studied empirically with both manipulative field experiments and long-term monitoring (Carpenter et al. 1992, 2001; Carpenter and Kitchell 1993; Carpenter 2003; Ludwig et al. 2003). This toy example is based on a minimal, three-column dataset, in which each row represents a set of measurements taken at a particular time. The first column is time, the second column is the level of the driver variable, such as the biomass of detritus or the concentration of critical nutrients, and the third column is the response variable, such as percent plant cover or $[O_2]$. In chapter 12, we evaluate the strength of these patterns in data from an enrichment experiment with the *S. purpurea*.

Multimodality in Frequencies of Response Variables   The first issue to consider is whether the system is one that can even be classified into alternative states. Perhaps with the exception of indicator species (e.g., Braun-Blanquet 1932; Siddig et al. 2016), communities are almost never classified on the basis of naturally discrete variables. Instead, we use variables measured on continuous scales and then bin them into "states" that may be biologically meaningful.

If we examine a system through time, the response variable should show evidence of bimodality (two alternative states) or multimodality (three or more alternative states). In other words, there should be some values or ranges of the response variable in which the system spends large periods of time, and other values that are improbable or transient (Hastings et al. 2018). In contrast, for the null hypothesis that a continuous response variable is simply tracking its continuous environmental driver, small changes in the driver variable should be accompanied by comparably small changes in the response variable. There should be no pronounced tendencies for some levels to be favored while others are unlikely.

Figure F.2 illustrates these idealized patterns for the frequency distribution of dissolved oxygen in an aquatic system with linear tracking (figure F.2, top) versus

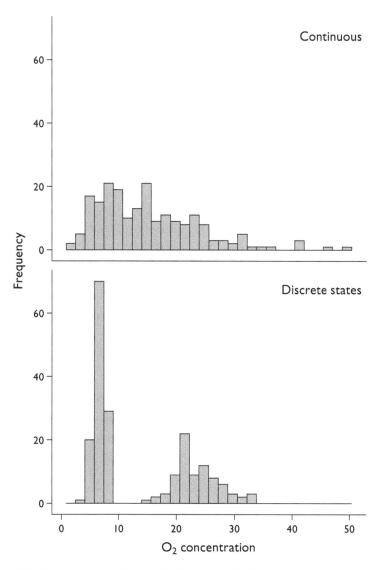

FIGURE F.2. Continuous versus discrete distributions of dissolved oxygen concentration ($[O_2]$) in a simulated aquatic ecosystem. **Top**: a continuous gamma distribution of $[O_2]$ values with a typically long right-hand tail ($\Gamma(3, 5.2)$). **Bottom**: a mixture of two gamma distributions with different means and variances (shape = 45, scale = 0.15; shape = 40, scale = 0.6), corresponding to a low $[O_2]$ "eutrophic state" and a high $[O_2]$ "oligotrophic state." Bimodality or multimodality in space or time of a response variable could be interpreted as evidence for multiple states.

one in which there is a bimodal distribution corresponding to oligotrophic and eutrophic states (figure F.2, bottom). Of course, this second pattern also could arise with environmental tracking, but it would naturally lead to the question of why the distribution of the driver variable is bimodal. Bi- or multimodality itself is a relatively weak pattern, but one that needs to first be established to justify the use of a model that bins continuous variation in nature into a smaller number of discrete states.

Rapid Nonlinear Change Between Alternative States   With incremental changes in a driver, most scenarios of alternative states predict a period of rapid nonlinear change from one to another equilibrium state, with stationary distributions of the response variable that differ in their mean levels between the two equilibria (figure F.3, bottom). In contrast, with simple environmental tracking, small changes in the driver variable will be accompanied by small changes in the state variable (figure F.3, top). The response variable may also lag behind temporal changes in the driver. But if time lags are present, they should not be any different for high and low levels of the driver. This distinction is especially important for understanding the initial collapse and long-term recovery of systems with slowly changing drivers.

Critical Slowing Down Preceding a State Change   Dynamic models of transitions between alternative states predict a period of "critical slowing down" (CSD) in which the variance or autocorrelation of the response variable increases as the system gets close to a tipping point (middle of main text figure 11.1; figures F.3, bottom and F.4; and Scheffer et al. 2009). To date, this increase in temporal variance has been detected in some highly simplified systems in laboratory experiments that may not be representative of complex multispecies assemblages (Drake and Griffen 2010; Veraart et al. 2012; Dai et al. 2013). More compelling ecological examples of CSD have been observed both in long-term observations and following experimental manipulations of entire lakes (Carpenter and Brock 2006; Brock and Carpenter 2010). Recent studies that have used space-for-time substitutions also have found evidence for CSD and changes in variance across aridity gradients in dry land ecosystems (Eby et al. 2017). But outside of lake ecosystems, experimental evidence for CSD remains limited, and the theoretical literature continues to grow faster than the data supporting it.

Moreover, increases in temporal variance and CSD do not appear in all systems undergoing regime shifts (Yasuhara et al. 2008; Hastings and Wysham 2010; Boettiger and Hastings 2012; Boettiger et al. 2013; Guttal et al. 2013; Doncaster et al. 2016). The utility of these indicators also is conditional on the size of the "moving window" in which the temporal variance is calculated relative to the speed of the response variable or the lifespan of the organism(s) responding to

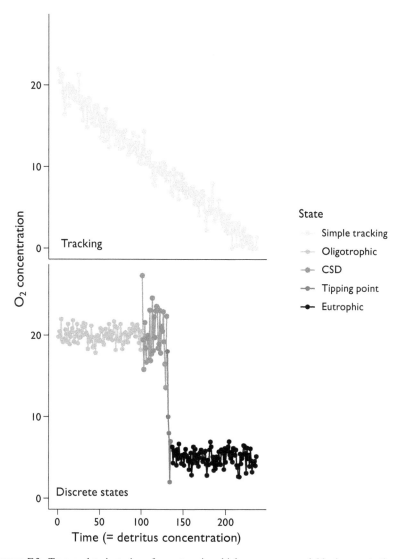

FIGURE F.3. Temporal trajectories of a system in which a response variable (concentration of dissolved oxygen) simply tracks a driver variable (detritus concentration) through time (**top** panel), versus a system with two alternate states (**bottom** panel). Grayscale indicates different dynamics: oligotrophic state, with high [O_2]; critical slowing down (CSD), with an increase in variance of the state variable; tipping point, with strong nonlinear transition between states; eutrophic state, with low [O_2]. In both trajectories, stochastic noise has been added.

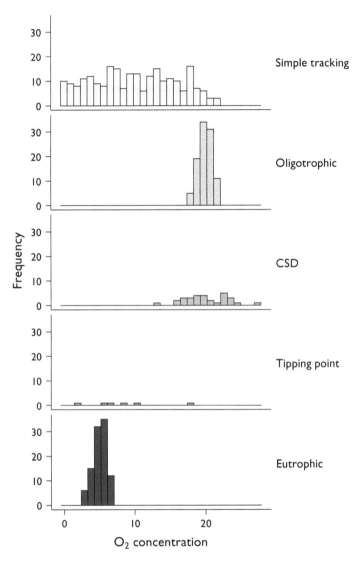

FIGURE F.4. Histograms of dissolved oxygen concentrations in a system with simple tracking (upper panel) versus a system with alternative states (four lower panels) as it progresses from a oligotrophc state, through critical slowing down (CSD) and a tipping point, and finally to a eutrophic state. Shaded levels as in the lower panel of figure F.3.

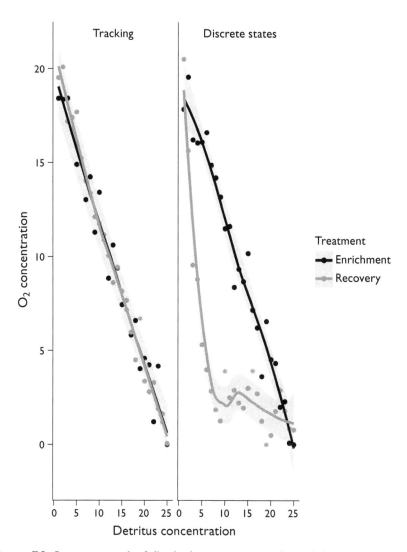

FIGURE F.5. State-space graph of dissolved oxygen concentration and detritus values in an enrichment experiment to test for hysteresis. Gray points and lines indicate times during which the system is enriched by a steady increase in detritus values. Black points and lines indicate times during which enrichment has ended, and the system is recovering. The **left** panel shows a system in which the response simply tracks the driver. The **right** panel shows hysteresis because the recovery of the system is unexpectedly slow after the driver has been reduced.

the underlying driver (Carpenter and Brock 2006; Yasuhara et al. 2008; Guttal et al. 2013). Ideally, the moving window should include serially autocorrelated observations across many lifespans of the responding organism (Yasuhara et al. 2008; Bestelmeyer et al. 2011).

Hysteresis Following Relaxation of the Driver Variable   Critical slowing down describes a signature pattern in temporal variance as a system approaches a tipping point. But once the tipping point has been crossed, there may be a long time lag before the system returns—if ever it does—to the original state. The asymmetry between a temporal pattern of rapid collapse before a transition and a sluggish return after the shift (hysteresis) also is a key signature of alternative states. However, it is difficult to interpret hysteresis without commensurate data on the underlying driver(s). If the level or concentration of the driver has progressively accumulated during the enrichment period, the asymmetry simply could reflect a constant rate of processing or conversion by the system, rather than a new stable state. Better evidence comes from state-space graphs, in which the temporal trajectory of both the driver and response are traced through time. The signature of hysteresis occurs when, during the return phase, the response of the system to the same level of the driver differs significantly between the enrichment and recovery phases of the trajectory (figure F.5). Once again, the best evidence for hysteresis in ecological systems has come from eutrophic systems in which water quality and $[O_2]$ show a hysteretic response to inputs of sediment, detritus, or limiting nutrients (Scheffer and Carpenter 2003; Middelburg and Levin 2009; Janse et al. 2010). The apparent "failure" of many restoration projects is encompassed in this graph because the system has not returned to its previous state, even though the suspected drivers have been reduced successfully to desired levels (Schindler 2012).

On Biodiversity, Ecosystem Function, and *omics

Chapter 13 synthesizes our work with tipping points and regime shifts in the context of biodiversity and ecosystem function. In this appendix, we fist define what we mean by ecosystem function. We then discuss why we think various types of "*omes" and "meta*omic" profiles may be useful indicators or measures to link biodiversity and ecosystem function. We close with a discussion of similarities and differences between sampling macrobial biodiversity and the diversity of *omes and meta*omes.

G.1 ECOSYSTEM SERVICES AND FUNCTIONS

Various typologies have been proposed to distinguish among ecosystem functions, services, and goods (de Groot et al. 2002), but most of these are inevitably variants of services that are provided to humans (Boyd and Banzhaf 2007). Whether it is specific goods that may profit an individual monetarily (e.g., increased biomass in an agricultural plot) or psychologically (e.g., sense of well-being, mental health), or general services that benefit society as a whole (e.g., climate control, soil retention), these functions all reflect the direct or indirect needs of humans.

These typologies also reflect the Clementsian "superorganism" view of an ecosystem as a set of integrated parts that work together (Clements 1916), and the idea that these functions reflect the "health" of the ecosystem (Callicott et al. 1999). But there is no agreement on what the important functions are or whether their specific rates are beneficial or detrimental to ecosystem health and ecosystem services (Schröter et al. 2014).

Perhaps a more useful framework is to consider an ecosystem from the cybernetics perspective: a set of modules with feedback loops through which "information" flows (Patten 1959; Müller 1997). An ecosystem can be represented as a network with pools and fluxes of nutrients, biomass, and energy (Patten 1959; Szyrmer and Ulanowicz 1987; Ulanowicz 2004). At the level of the individual organism, there are a relatively small number of physiological processes of plants and animals that ultimately drive these flows: water and nutrient uptake, photosynthesis, consumption, metabolism, excretion, sequestration, and decomposition.

But how can we extract a concept of information from these processes? One perspective is that the individuals of different species are the units of information, so that a species-rich community with a relatively even abundance distribution has more information (and enhanced ecosystem function) than a species-poor community in which a few species dominate the biomass spectrum (Wittebolle et al. 2009; cf. Patten 1959). This formulation leads naturally to the connection between species diversity and ecosystem functions.

G.2 BIODIVERSITY AND ECOSYSTEM FUNCTION

A major research focus in ecology since the mid-1990s has been to identify connections, and quantify the strengths and directions of relationships, between biodiversity and ecosystem function (Hooper et al. 2005). Much experimental work, mostly at small spatial scales in temperate-zone grasslands, has established that processes such as biomass accumulation and nutrient retention are indeed higher for multispecies assemblages than for monocultures (Balvanera et al. 2006). However, it has been difficult to tease apart how much of this effect reflects changes in species number and how much results from changes in species composition (Huston 1997; Cardinale et al. 2007; Pillai and Gouhier 2019). The strength of the biodiversity effect also is rarely compared to the strength of other drivers of ecosystem function and environmental change (Maestre et al. 2012; van der Plas 2019).

The abundance, richness, and composition of species are both causes and consequences of ecosystem processes and are themselves dynamic and changeable. Should these species be the units of information in a cybernetic analysis of ecosystems? Although some species, including keystone and foundation species, play critical roles in determining relative abundance of co-occurring species and overall ecosystem structure and dynamics (Ellison 2019), many important ecosystem functions can change even when species richness and composition are held constant (Hillebrand et al. 2008). Conversely, some species are functionally redundant, so their presence or absence has only a weak effect on ecosystem processes (Walker 1992). It also increasingly is clear that intraspecific variation—individuals of different genotypes, ages, sizes, or phenotypes—contribute to variation in ecosystem function (Reusch and Wood 2007; Violle et al. 2012).

G.2.1 Elements, Proteins, and *omes

In contrast to the dynamics of species assemblages, matter and elemental composition are conserved in closed systems. Indeed, elemental analysis forms the basis for studies of nutrient stoichiometry, in which we consider organisms simply as mixtures of different elements (see chapters 2–4 and appendix B). However, at the level of an entire ecosystem, elemental analysis may be too simpified to reflect ecosystem processes. Even for relatively simple molecules, the same element can be biologically inert (N_2) or be packaged in a form that is biologically highly reactive (NH_4, NO_3).

A more useful chemical unit than the element might be the protein, which is a relatively large biomolecule consisting of chains of amino acids that fold into a functional three-dimensional structure (Fersht 2017). Proteins form important structural components of organisms and perform a variety of biological functions (Somero et al. 2017). They are rapidly synthesized and degraded during all cellular processes, and the amino-acid chains that comprise them are synthesized directly from information encoded in DNA sequences.

Because proteins behave as enzymes for important biological reactions, they seem closer to the functioning of ecosystems (however that is operationally defined) than either the whole organism (which carries out a variety of functions) or the underlying DNA sequences (which are not decoded until particular proteins are needed; Gotelli et al. 2012). Moreover, proteins identified in a particular system can be screened against libraries of proteins from previously studied organisms, providing not only their identity but also a mapping to broad classes of metabolic pathways (Kanehisa et al. 2016).

In most ecosystems, it would not be practical to sequence all the proteins that are present because most of them are bound up in the bodies of constituent organisms. And until recently, it was too impractical and inefficient to sequence and identify unknown proteins that were not part of a small number of well-studied model systems (e.g., *Drosophila melanogaster, Caenorhabditis elegans, Mus musculus* or *Arabidopsis thaliana*; chapter 1, §1.5). But recent advances in proteomics now make it feasible to consider surveys of all the proteins in an ecosystem as a way to quantify its functions (Schneider and Riedel 2010). In particular, we might speculate that a healthy ecosystem is one that has a greater number and diversity of biologically active proteins that should allow for better homeostatic control in variable environments and provide more products and services that are important to humans.

Of course, metaproteomics—the identification and classification of heterogeneous mixes of proteins sampled from an entire (eco)system—is just one of many possible meta*omes that could be used to characterize ecosystem function: metagenomics, metatranscriptomics, and metametabolomics are other ways to sequence, analyze, or screen large assemblages of candidate molecules. Although metagenomics and metatranscriptomics do contain the genetic information and transcript encodings of proteins (Vandenkoornhuyse et al. 2010), these profiles are poorly correlated with the abundance and diversity of proteins that are actually expressed by organisms in an ecosystem (Vogel and Marcotte 2012). Metametabolomics provides a snapshot of the production of a small number of relatively simple metabolic products (Bundy et al. 2009), but, like nutrient stoichiometry, oversimplifies the array of biomolecules synthesized by a functioning ecosystem. Thus, on functional grounds, we favor metaproteomics as the best proxy variable for a general assay of ecosystem function.

G.2.2 Sampling Bio- and *omic Diversity: Similarities

There are many similarities between a meta*omics survey and a traditional biodiversity survey. First, the survey has to be both taxonomically restricted and physically bounded within a specified area or habitat (such as the butterflies of tropical forest light gaps; Hill et al. [2001] or the microbiome of household washing machines Callewaert et al. [2015]). Second, the biologically packaged units (individuals, genes, transcripts, polypeptides) must be randomly sampled. However, this sample will always represent only a tiny fraction of the material that would be required to capture all of the diversity in the system (Hughes et al. 2001).

Next, each individual element has to be identified and classified into one of a set of nonoverlapping "bins" (species, OTUs, proteins). The number of bins, composition, and relative abundances of this sorted distribution are then summarized with one or more biodiversity metrics, such as the familiar number of species or indices of evenness or species composition (Magurran and McGill 2011).

Finally, these diversity metrics need to be evaluated statistically for replicated comparisons of different experimental treatments (such as controls versus nutrient enrichment) or classification groups (such as the microbiomes of healthy versus diseased individuals). Because biodiversity and meta*omic surveys are always characterized by a few common elements and many rare ones (Morlon et al. 2009), these biodiversity statistics are sensitive to the number of units that are sampled and the specific methods that are used to sample them (Gotelli and Colwell 2001). Extrapolation and interpolation methods for diversity statistics are now well developed and allow for standardized comparisons of any level of diversity and whenever sampling effort or abundance is not equivalent (Chao et al. 2014).

G.2.3 Sampling Bio- and *omic Diversity: Differences

There also are some important differences between meta*omic surveys and traditional biodiversity surveys. In a biodiversity survey, we rely on taxonomic keys to distinguish the different species (Farnsworth et al. 2013). In a metaproteomic survey, the protein sequences are mapped back to sequence libraries (NCBI Resource Coordinators 2017). Because these libraries are mostly based on classical model organisms studied in a biomedical context, proteins collected in environmental surveys may be misclassified, or simply cannot be classified at all. For example, in our metaproteomic surveys of control and prey-supplemented *S. purpurea* microecosystems (described in chapter 13), we mapped the 220 most common protein sequences in each treatment to taxonomic groups by running a BLAST (Basic Local Alignment Search Tool) homology search on the metagenomic sequence data to reveal corresponding to the protein hits (Altschul et al. 1997). Thirteen percent of the enriched proteins and 21% of the control proteins could not be mapped back to a taxonomic group in the BLAST library.

Although both sampling for species and meta*omic diversity suffer from undersampling, the problem is far more severe for proteomic surveys. Estimates of metaproteomic diversity in microbial ecosystems range from 10^4 to 10^9 expressed proteins (Wilmes and Bond 2006). There are approximately 6000–60,000 protein-encoding genes in the genome of a single prokaryote or eukaryote species (Van Straalen and Roelofs 2012). Even a single tissue sampled from a single multicellular species contains hundreds or thousands of distinct proteins (Shen et al. 2004). In ecological assemblages that contain both microbes and macrobes, diversity will probably exceed 10^{10} proteins (Gotelli et al. 2012).

Further, the proteome of an individual organism, in contrast to its genome, does not remain constant through its lifetime because of ontogenetic changes (Guercio et al. 2006) and responses to biotic and abiotic conditions (Nesatyy and Suter 2007). Although there is added complexity in working with and analyzing proteomes, they do appear to be a better reflection of the "molecular phenotype" of the individual (Biron et al. 2006b). The metaproteome also may be a better indicator of changes in ecosystem function, as it can change even if species diversity does not.

Moreover, unlike individuals of a given species, proteins are not sampled and identified as intact objects. Rather, before proteins can be sequenced and identified, they are proteolytically digested and cut into smaller polypeptide fragments (Henzel et al. 1993). These fragments are then sequenced and reassembled to identify the individual proteins.

A familiar analogy for an assemblage of species is a large jar of multicolored jelly beans (Longino et al. 2002). An investigator samples its "biodiversity" by taking a handful of jelly beans, sorting and counting the colors, and then uses a statistical inference to estimate the diversity of jelly beans in the entire jar. In contrast, the "jar of proteins" is filled with hard candies. After an investigator draws a handful of candies, they are shattered with a hammer, and the investigator counts and re-assembles the individual shards (i.e., polypeptides) before estimating the diversity or proteins in the sample. This additional complicating step means that the uncertainty in identification of proteins increases as polypeptide sequences get smaller and rarer. For this reason, most studies, including ours, emphasize changes in the most common protein varieties in a treatment.

One final sampling issue is that, in meta*omic surveys, it has been common practice to pool replicate samples from an area or experimental treatment to provide more material, which in turn yields more accurate quantification of the relative abundance of proteins. However, if there is heterogeneity in protein profiles among the replicates within a treatment, this pooling will tend to exaggerate the statistical differences that are reported between treatments (Karp et al. 2005). "Technical replicates" often are created by subsampling from the pooled replicates, but this creates a problem of pseudoreplication (Hurlbert 1984). As in ecological studies, principles of randomization, replication, and statistical independence need to be heeded in *omics research (Biron et al. 2006a). In the results presented in chapter 13, we are mostly describing patterns from pooled data, but these have been confirmed with additional parametric statistical tests and randomization analyses of unpooled data.

G.3 DIVERSITY AND INFORMATION CONTENT IN *OMES

One way to quantify diversity in a (meta)proteome is to consider it as a cybernetic system (§G.1) and measure its information content. To do this, we asked, if we were to randomly grab two protein molecules from a pitcher, what would be the probability that they are different proteins? We used Hurlbert's (1971) PIE (probability of an interspecific encounter) to answer this question. As a single measure of information to characterize a rank-abundance distribution, PIE has several advantages. First, it is relatively insensitive to sampling variation (Gotelli and Graves 1996) and is influenced most strongly by the relative abundance distribution of the most common proteins. Second, it is algebraically related to the Gini coefficient of income inequality (Wittebolle et al. 2009). Third, it is a measure of the slope of an individual-based rarefaction curve at is base (Olszewski 2004) and can be transformed to one of the Hill numbers ($q = 2$) of biodiversity (Chao et al. 2014).

Notes

CHAPTER 1. INTRODUCTION: WHY SCALE?

This chapter is modified and expanded from Ellison (2018).

1. In this world, there are both large-scale macrocosms and small-scale microcosms; the latter are at least as important as the former.

2. viz. God never meant that man should scale the heav'ns / By strides of human wisdom (Cowper 1785: iii, 221).

CHAPTER 6. THE SMALL WORLD: DEMOGRAPHY OF A LONG-LIVED PERENNIAL CARNIVOROUS PLANT

1. It is uncertain that everything is certain.

2. See appendix C, equation C.3 for definition and calculation of elasticities.

3. The Dirichlet distribution with parameters $\{\alpha_i\}_{i=1}^K$ is the discrete probability distribution of α_i successful outcomes of a random variable x_i in a sequence of $\Sigma_{j=1}^K \alpha_j$ independent experiments, each with a probability of $\alpha_j / \Sigma_{j=1}^K$. For a demographic example, if a *Sarracenia* cohort of 10 juvenile plants is followed for one generation, some (e.g., 5) of these 10 can either stay as juveniles, while others transition into the non-flowering adult stage (3) or die (2). The point estimates of the probabilities of these mutually exclusive events are $\{0.5, 0.3, 0.2\}$. In contrast, using a Dirichlet sampler [e.g., the rdirichlet function in the gtools R package of Warnes et al. 2008]) would give the full probability distribution for each of these point estimates.

CHAPTER 7. SCALING UP: INCORPORATING DEMOGRAPHY AND EXTINCTION RISK INTO SPECIES DISTRIBUTION MODELS

1. See appendix D for an introduction to basics of MaxEnt and other SDMs.

2. https://www.gbif.org/

3. https://nadp.slh.wisc.edu/data/

4. In fact, as we completed this book in spring 2020, the US Environmental Protection Agency released its final rule rolling back earlier increases in fuel-economy standards for passenger cars and light trucks. See Federal Dockets NHTSA-2018-0067 and EPA-HQ-OAR-2018-0283; and online text of the final rule at https://www.epa.gov/sites/production/files/2020-03/documents/final-fr-safe-preamble-033020.pdf. The roll-back is being challenged in the courts.

CHAPTER 8. CONTEXT: COMMUNITY ECOLOGY, COMMUNITY ECOLOGIES, AND COMMUNITIES OFF ECOLOGISTS

1. We discuss this intellectual history in more detail in appendix E.

2. The *Oxford English Dictionary* defines a "community" as a body of people or things viewed collectively (OED Online 2018a). By extension, an ecological community is defined therein as "a group of animals or plants in the same place; [or] a group of organisms growing or living together in natural conditions or occupying a specified area." Animals receive historical priority here because the first occurrence of "community" in an ecological sense recorded by the OED—the epigraph of this chapter—is by Hervey (1747).

CHAPTER 9. THE SMALL WORLD: STRUCTURE AND DYNAMICS OF INQUILINE FOOD WEBS IN *SARRACENIA PURPUREA*

1. See appendix E, table E.4 for definitions and calculations of measures and metrics of food-web structure.

CHAPTER 10. SCALING UP: THE GENERALITY OF THE *SARRACENIA* FOOD WEB AND ITS VALUE AS A MODEL EXPERIMENTAL SYSTEM

1. It awakens in you the idea of the infinite; you can see in it, as in a microcosm, the details and the instability of a universe changing constantly before our very eyes.

2. https://www.globalwebdb.com

3. https://digitalcommons.rockefeller.edu/cohen_joel_laboratory/1/

4. https://www.nceas.ucsb.edu/interactionweb/resources.html

5. Composite web data kindly provided to us by Olivier Dézerald.

6. Note that a biogeochemical web can be transformed to a trophic web by replacing entries with ones. All of the metrics that we computed are based on such binary interaction matrices.

CHAPTER 11. CONTEXT: TIPPING POINTS AND REGIME SHIFTS

1. All this is for the best; if there is a volcano in Lisbon, it could not be elsewhere; it is impossible that things are not where they are, because everything is good.

2. Data from a search done using Web of Science on June 17, 2019 with the search string ("regime shift" OR "regime change" OR "regime" OR "tipping point").

3. Note that critical drivers need not appear as explicit terms in mathematical models of tipping points and regime shifts. Indeed, if the model does not include a critical driver, its behavior often can be mimicked by including in its place a time lag for unspecified forces. For example, the population cycles predicted by the classic Lotka–Volterra predator–prey model can be mimicked superficially by a single-species model of logistic growth with a time lag (May and McLean 2007). However, the dynamics of the two systems are distinctly different after a perturbation is introduced.

CHAPTER 12. THE SMALL WORLD: TIPPING POINTS AND REGIME SHIFTS IN THE *SARRACENIA* MICROECOSYSTEM

1. For he was stirring up billows in a ladle. The presumed origin of the phrase "tempest in a teapot."

CHAPTER 13. SCALING UP: USING *OMICS TO IDENTIFY ECOSYSTEM STATES AND TRANSITIONS

1. https://en.wikipedia.org/wiki/Tricorder
2. See Appendix G, §G.1 for conceptual discussion and computational details.

CHAPTER 14. CONCLUSION: WHITHER *SARRACENIA*?

1. Whether such states are truly "stable," merely "resilient" (Thorén and Olsson 2017), or ephemerally "transient" (Hastings et al. 2018) depends on the precise definition of stability (of which ecologists have literally hundreds: Grimm and Wissell 1997; Donohue et al. 2016). Rather than wade into that jargon-laded quagmire of assumptions and metaphors, we simply let the experimental data speak for themselves.

2. This interpretation guides most of restoration ecology, which explicitly aims to recreate conditions prior to human intervention (usually European, *ca.* 1491), which have been wistfully described as "ecologically privileged assemblages of organisms, endowed with distinctive qualities of stability, beauty, and self-organizing capacity" (Jordan and Lubick 2011, p. 2).

APPENDIX A. THE NATURAL HISTORY OF *SARRACENIA* AND ITS MICROECOSYSTEM

1. Ironically, the Asian pitcher plant *Nepenthes aristolochiodes* Jebb & Cheek was named for the resemblance of its pitchers to the flowers of *Aristolochia* (Jebb and Cheek 1997). Others have mistakenly identified *Aristolochia* as carnivorous because many species are pollinated by flies that are lured to the flowers by secondary compounds that stink like dead and decomposing insects. But *Aristolochia* does not kill its pollinators. Rather, the flies are led to the hidden anthers and stigmas by downward pointing hairs (similar to those that discourage captured prey from escaping pitcher plants), temporarily held within the flower, and then released after pollination is complete (Oelschlägel et al. 2009, 2015; González and Pabón-Mora 2015).

2. Available online through the Cambridge Digital Library: https://cudl.lib.cam.ac.uk

3. In 2006, Stephanie Day, then an undergraduate student at Howard University, worked with us to digitize much of Frank Morton Jones's writings and unpublished data that were in the archives of the Department of Entomology at Yale University's Peabody Museum. This small digital archive is available at https://harvardforest.fas.harvard.edu/ellison/current-research/Frank-Morton-Jones.

APPENDIX C. DETERMINISTIC STAGE-BASED MODELS

1. The corresponding integral projection model for equation C.1 would be $n(y, t + 1) = \int_L^U K(y, x)n(x, t)dx$, where $n(x, t)dx$ is the number of individuals (n) with state variable x (e.g., size) $\in [x, x + dx]$. $K(y, x) = P(y, x) + F(y, x)$ is the projection kernel (replacing the projection matrix A in equation C.1); $P(\cdot)$ represents survival and growth and $F(\cdot)$ represents fecundity in all possible states between L and U. Like equation C.1, the integral projection model generates a stable age distribution and exponential population growth, and can be analyzed in much the same way as the simpler discrete-time matrix model. See Ellner and Rees (2006) and White et al. (2016) for additional details.

APPENDIX D. THE BASICS OF SPECIES DISTRIBUTION MODELS

1. https://www.gbif.org/
2. https://www.inaturalist.org/
3. Species occurrence data from a variety of portals can be accessed using the occ function in the spocc package in R (Chamberlain et al. 2020).

APPENDIX F. ON TIPPING POINTS AND REGIME SHIFTS

1. Keeling, with support from the International Geophysical Year program, began measuring [CO_2] at a time when he and his colleagues were already concerned that it was already close to causing a planetary tipping point.

References

Åberg, P. 1992. Size-based demography of the seaweed *Ascophyllum nodosum* in stochastic environments. Ecology **73**:1488–1501.

Abrahamson, W. G., and A. E. Weis. 1997. Evolutionary Ecology Across Three Trophic Levels: Goldenrods, Gallmakers, and Natural Enemies. Princeton University Press, Princeton, NJ.

Abril, A. B., and E. H. Bucher. 2009. A comparison of nutrient sources of the epiphyte *Tillandsia capillaris* attached to trees and cables in Cordoba, Argentina. Journal of Arid Environments **73**:393–395.

Ackerly, D., S. Dudley, S. Sultan, J. Schmitt, J. Coleman, C. Linder, et al. 2000. The evolution of plant ecophysiological traits: recent advances and future directions. BioScience **50**:979–995.

Adamec, L. 1997. Mineral nutrition of carnivorous plants: a review. Botanical Review **63**:273–299.

Adams, D. 1980. The Hitchhiker's Guide to the Galaxy. Harmony Books, New York.

Addicott, J. F. 1974. Predation and prey community structure: an experimental study of the effect of mosquito larvae on the protozoan communities of pitcher plants. Ecology **55**:475–492.

Adlassnig, W., E. Mayer, M. Peroutka, W. Pois, and I. K. Lichtscheidl. 2010. Two American *Sarracenia* species as neophyta in central Europe. Phyton **49**:279–292.

Adlassnig, W., G. Steinhauser, M. Peroutka, A. Musilek, J. H. Sterba, I. K. Lichtscheidl, and M. Bichler. 2009. Expanding the menu for carnivorous plants: uptake of potassium, iron and manganese by carnivorous pitcher plants. Applied Radiation and Isotopes **67**:2117–2122.

Aerts, R., and F. S. Chapin. 2000. The mineral nutrition of wild plants revisited: a re-evaluation of processes and patterns. Advances in Ecological Research **30**:1–67.

Agosta, S. 2006. On ecological fitting, plant–insect associations, herbivore host shifts, and host plant selection. Oikos **114**:556–565.

Ågren, G. 2008. Stoichiometry and nutrition of plant growth in natural communities. Annual Review of Ecology, Evolution, and Systematics **39**:153–170.

Ahn, H., and R. T. James. 2001. Variability, uncertainty, and sensitivity of phosphorus deposition load estimates in South Florida. Water, Air, and Soil Pollution **125**: 37–51.

Allen, S., A. Carlisle, E. White, and C. Evans. 1968. The plant nutrient content of rainwater. Journal of Ecology **56**:497–504.

Allesina, S., and M. Pascual. 2008. Network structure, predator-prey modules, and stability in large food webs. Theoretical Ecology **1**:55–64.

Allesina, S., and S. Tang. 2012. Stability criteria for complex ecosystems. Nature **483**:205–208.

Alley, R. B., P. U. Clark, P. Huybrechts, and I. Joughin. 2005. Ice-sheet and sea-level changes. Science **310**:456–460.

Allstadt, A. J., S. J. Vavrus, P. J. Heglund, A. M. Pidgeon, W. E. Thogmartin, and V. C. Radeloff. 2015. Spring plant phenology and false springs in the coterminous US during the 21st century. Environmental Research Letters **10**:104008.

Altschul, S. F., T. L. Madden, A. A. Schaffer, J. H. Zhang, Z. Zhang, W. Miller, and D. J. Lipman. 1997. Gapped BLAST and PSI-BLAST: a new generation of protein database search programs. Nucleic Acids Research **25**:3389–3402.

Ames, M. E. P. 1880. Notes on the flora of Plumas County. The California Horticulturalist and Floral Magazine **10**:325–329.

An, C. B., L. Y. Tang, L. Barton, and F. H. Chen. 2005. Climate change and cultural response around 4000 cal yr BP in the western part of Chinese Loess Plateau. Quaternary Research **63**:347–352.

Anderson, J., D. Inouye, A. McKinney, R. Colautti, and T. Mitchell-Olds. 2012a. Phenotypic plasticity and adaptive evolution contribute to advancing flowering phenology in response to climate change. Proceedings of the Royal Society, B: Biological Sciences **279**:3843–3852.

Anderson, J., A. Panetta, and T. Mitchell-Olds. 2012b. Evolutionary and ecological responses to anthropogenic climate change. Plant Physiology **160**:1728–1740.

Andersson, K., M. Tighe, C. Guppy, P. Milham, T. McLaren, C. Schefe, and E. Lombi. 2016. XANES demonstrates the release of calcium phosphates from alkaline vertisols to moderately acidified solution. Environmental Science & Technology **50**:4229–4237.

Angeler, D. G., C. R. Allen, C. Barichievy, T. Eason, A. S. Garmestani, N. A. J. Graham, et al. 2016. Management applications of discontinuity theory. Journal of Applied Ecology **53**:688–698.

Arber, A. 1921. The leaf structure of the Iridaceae, considered in relation to the phyllode theory. Annals of Botany **35**:301–336.

Arber, A. 1941. On the morphology of the pitcher-leaves in *Heliamphora, Sarracenia, Darlingtonia, Cephalotus,* and *Nepenthes.* Annals of Botany **5**:563–578.

Arita, H. T. 2016. Species co-occurrence analysis: pairwise versus matrix-level approaches. Global Ecology and Biogeography **25**:1397–1400.

Arita, H. T. 2017. Multisite and multispecies measures of overlap, co-occurrence, and co-diversity. Ecography **40**:709–718.

Armitage, D. W. 2016a. Bacteria facilitate prey retention by the pitcher plant *Darlingtonia californica.* Biology Letters **12**:20160577.

Armitage, D. W. 2016b. Time-variant species pools shape competitive dynamics and biodiversity-ecosystem function relationships. Proceedings of the Royal Society B: Biological Sciences **283**:20161437.

Armitage, D. W. 2017. Linking the development and functioning of a carnivorous pitcher plant's microbial digestive community. ISME Journal **11**:2439–2451.

Atwater, D. Z., J. L. Butler, and A. M. Ellison. 2006. Spatial distribution and impacts of moth herbivory on northern pitcher plants. Northeastern Naturalist **13**:43–56.

Atwood, M. 2004. Our cat goes to heaven. Brick **73**:8–9.

Augspurger, C. K. 2013. Reconstructing patterns of temperature, phenology, and frost damage over 124 years: spring damage risk is increasing. Ecology **94**:41–50.

Bachman, S., J. Moat, A. W. Hill, J. de la Torre, and B. Scott. 2011. Supporting Red List threat assessments with GeoCAT: geospatial conservation assessment tool. ZooKeys **150**:117–126.

Baillon, H. 1870. Sur le developpement des feuilles des *Sarracenia*. Comptes Rendus de l'Académie des Sciences **71**:331–333.

Baird, D., and R. E. Ulanowicz. 1989. The seasonal dynamics of the Chesapeake Bay ecosystem. Ecological Monographs **59**:329–364.

Baiser, B., R. Ardeshiri, and A. M. Ellison. 2011. Species richness and trophic diversity increase decomposition in a co-evolved food web. PLoS ONE **6**:e20672.

Baiser, B., H. L. Buckley, N. J. Gotelli, and A. M. Ellison. 2013. Predicting food-web structure with metacommunity models. Oikos **122**:492–506.

Baiser, B., N. J. Gotelli, H. L. Buckley, T. E. Miller, and A. M. Ellison. 2012. Geographic variation in network structure of a Nearctic aquatic food web. Global Ecology and Biogeography **21**:579–581.

Balvanera, P., A. B. Pfisterer, N. Buchmann, J.-S. He, T. Nakashizuka, D. Raffaelli, and B. Schmid. 2006. Quantifying the evidence for biodiversity effects on ecosystem functioning and services. Ecology Letters **9**:1146–1156.

Banasek-Richter, C., L. F. Bersier, M. F. Cattin, R. Baltensperger, J. P. Gabriel, Y. Merz, et al. 2009. Complexity in quantitative food webs. Ecology **90**:1470–1477.

Barnosky, A. D., J. H. Brown, G. C. Daily, R. Dirzo, A. H. Ehrlich, P. R. Ehrlich, et al. 2013. Scientific consensus on maintaining humanity's life support systems in the 21st century: information for policy makers. http://mahb.stanford.edu/consensus-statement-from-global-scientists/.

Barnosky, A. D., P. R. Ehrlich, and E. A. Hadly. 2016. Avoiding collapse: grand challenges for science and society to solve by 2050. Elementa **4**:1–9.

Barnosky, A. D., E. A. Hadly, J. Bascompte, E. L. Berlow, J. H. Brown, M. Fortelius, et al. 2012. Approaching a state shift in Earth's biosphere. Nature **486**:52–58.

Barnosky, A. D., N. Matzke, S. Tomiya, G. O. U. Wogan, B. Swartz, T. B. Quental, et al. 2011. Has the Earth's sixth mass extinction already arrived? Nature **471**:51–57.

Barry, K. E., G. A. Pinter, J. W. Strini, K. Yang, I. G. Lauko, S. A. Schnitzer, et al. 2019. A universal scaling method for biodiversity-ecosystem functioning relationships. bioRxiv doi:10.1101/662783.

Bartlett, A. A., 1969. Arithmetic, population, and energy. https://www.youtube.com/watch?v=sI1C9DyIi_8.

Bartram, W. 1793. Travels Through North & South Carolina, Georgia, East & West Florida, the Cherokee Country, the Extensive Territories of the Muscogulges, or Creek Confederacy, and the Country of the Chactaws; Containing an Account of the Soil and Natural Productions of Those Regions, Together with Observations on the Manners of the Indians. James & Johnson, Philadelphia.

Bauer, U., H. F. Bohn, and W. Federle. 2008. Harmless nectar source or deadly trap: *Nepenthes* pitchers are activated by rain, condensation and nectar. Proceedings of the Royal Society B: Biological Sciences **275**:259–265.

Bauer, U., R. Jetter, and S. Poppinga, 2018. Non-motile traps. In A. M. Ellison and L. Adamec, editors. Carnivorous Plants: Physiology, Ecology, and Evolution, pp. 194–206. Oxford University Press, Oxford.

Beisner, B. E., D. T. Haydon, and K. Cuddington. 2003. Alternative stable states in ecology. Frontiers in Ecology and the Environment **1**:376–382.

Bender, E. A., T. J. Case, and M. Gilpin. 1984. Perturbation experiments in community ecology: theory and practice. Ecology **65**:1–13.

Benitez-Nelson, C. 2000. The biogeochemical cycling of phosphorus in marine systems. Earth Science Reviews **51**:109–135.

Bennett, K. T., and A. M. Ellison. 2009. Nectar, not colour, may lure insects to their death. Biology Letters **5**:469–472.

Berendse, F., N. Van Breemen, H. Rydin, A. Buttler, M. Heijmans, M. R. Hoosbeek, et al. 2001. Raised atmospheric CO_2 levels and increased N deposition cause shifts in plant species composition and production in *Sphagnum* bogs. Global Change Biology **7**:591–598.

Berlow, E. L., S. A. Navarrete, C. J. Briggs, M. E. Power, and B. A. Menge. 1999. Quantifying variation in the strengths of species interactions. Ecology **80**:2206–2224.

Bestelmeyer, B. T., A. M. Ellison, W. R. Fraser, K. B. Gorman, S. J. Holbrook, C. M. Laney, et al. 2011. Analysis of abrupt transitions in ecological systems. Ecosphere **2**:129.

Bierman, P. R., P. T. Davis, L. B. Corbett, N. A. Lifton, and R. C. Finkel. 2015. Cold-based Laurentide ice covered New England's highest summits during the Last Glacial Maximum. Geology **43**:1059–1062.

Biggs, R., S. R. Carpenter, and W. A. Brock. 2009. Turning back from the brink: detecting an impending regime shift in time to avert it. Proceedings of the National Academy of Sciences of the USA **106**:826–831.

Bird, H. 1903. New histories and species in *Papaipema* and Hydrœcia. The Canadian Entomologist **35**:91–94.

Biron, D. G., C. Brun, T. Lefevre, C. Lebarbenchon, H. D. Loxdale, F. Chevenet, et al. 2006a. The pitfalls of proteomics experiments without the correct use of bioinformatics tools. Proteomics **6**:5577–5596.

Biron, D. G., H. D. Loxdale, F. Ponton, H. Moura, L. Marché, C. Brugidou, and F. Thomas. 2006b. Population proteomics: an emerging discipline to study metapopulation ecology. Proteomics **6**:1712–1715.

Bittleston, L. S., M. Gralka, G. E. Leventhal, I. Mizrahi, and O. X. Cordsero. 2020. Context-dependent dynamics lead to the assembly of functionally distinct microbial communities. Nature Communications **11**:1440.

Blanchet, F. G., K. Cazelles, and D. Gravel. 2020. Co-occurrence is not evidence of ecological interactions. Ecology Letters **23**:1050–1063.

Błędzki, L. A., and A. M. Ellison. 1998. Population growth and production of *Habrotrocha rosa* Donner (Rotifera: Bdelloidea) and its contribution to the nutrient supply of its host, the northern pitcher plant, *Sarracenia purpurea* L. (Sarraceniaceae). Hydrobiologia **385**:193–200.

Błędzki, L. A., and A. M. Ellison. 2003. Diversity of rotifers from northeastern USA bogs with new species records for North America and New England. Hydrobiologia **497**:53–62.

Blois, J. L., J. W. Williams, M. C. Fitzpatrick, S. T. Jackson, and S. Ferrier. 2013. Space can substitute for time in predicting climate-change effects on biodiversity. Proceedings of the National Academy of Sciences of the USA **110**:9374–9379.

Blüthgen, N. 2010. Why network analysis is often disconnected from community ecology: a critique and an ecologist's guide. Basic and Applied Ecology **11**:185–195.

Boettiger, C., and A. Hastings. 2012. Early warning signals and the prosecutor's fallacy. Proceedings of The Royal Society B: Biological Sciences **279**:4734–4739.

Boettiger, C., N. Ross, and A. Hastings. 2013. Early warning signals: the charted and uncharted territories. Theoretical Ecology **6**:255–264.

Bohn, H. F., and W. Federle. 2004. Insect aquaplaning: *Nepenthes* pitcher plants capture prey with the peristome, a fully wettable water-lubricated anisotropic surface. Proceedings of the National Academy of Sciences of the USA **101**:14138–14143.

Borrett, S. R., and M. K. Lau. 2014. enaR: an R package for ecosystem network analysis. Methods in Ecology and Evolution **11**:1206–1213.

Botkin, D. B., H. Saxe, M. B. Araújo, R. Betts, R. H. W. Bradshaw, T. Cedhagen, et al. 2007. Forecasting the effects of global warming on biodiversity. BioScience **57**:227–236.

Bott, T., G. A. Meyer, and E. B. Young. 2008. Nutrient limitation and morphological plasticity of the carnivorous pitcher plant *Sarracenia purpurea* in contrasting wetland environments. New Phytologist **180**:631–641.

Boyd, J., and S. Banzhaf. 2007. What are ecosystem services? The need for standardized environmental accounting units. Ecological Economics **63**:616–626.

Boynton, P., C. N. Peterson, and A. Pringle. 2019. Superior dispersal ability can lead to persistent ecological dominance throughout succession. Applied and Environmental Microbiology **85**:e02421–18.

Boynton, W., J. H. Garber, R. Summers, and W. Kemp. 1995. Inputs, transformations and transport of nitrogen and phosphorus in Chesapeake Bay and selected tributaries. Estuaries **18**:285–314.

Bradshaw, W. E., 1983. Interaction between the mosquito *Wyeomyia smithii*, the midge *Metriocnemus knabi*, and their carnivorous host *Sarracenia purpurea*. J. H. Frank and L. P. Lounibos, editors. Phytotelmata: Terrestrial Plants as Hosts for Aquatic Insect Communities, pp. 161–189. Plexus Press, Medford, NJ.

Bradshaw, W. E., and R. A. Creelman. 1984. Mutualism between the carnivorous pitcher purple pitcher plant and its inhabitants. American Midland Naturalist **112**:294–304.

Bradshaw, W. E., and C. M. Holzapfel. 2001. Genetic shift in photoperiodic response correlated with global warming. Proceedings of the National Academy of Sciences of the USA **98**:14509–14511.

Braun-Blanquet, J. 1932. Plant Sociology: The Study of Plant Communities. McGraw-Hill, New York.

Brewer, S. J. 2019. Inter- and intraspecific competition and shade avoidance in the carnivorous pale pitcher plant in a nutrient-poor savanna. American Journal of Botany **106**:81–89.

Brock, W. A., and S. R. Carpenter. 2010. Interacting regime shifts in ecosystems: implication for early warnings. Ecological Monographs **80**:353–367.

Brook, B. W., E. C. Ellis, and J. C. Buettel, 2018. What is the evidence for planetary tipping points? In P. Kareiva, M. Marvier, and B. Silliman, editors. Effective Conservation Science: Data not Dogma, pp 51–57. Oxford University Press, Oxford.

Brook, B. W., E. C. Ellis, M. P. Perring, A. W. Mackay, and L. Blomqvist. 2013. Does the terrestrial biosphere have planetary tipping points? Trends in Ecology & Evolution **28**:396–401.

Brower, J. H., and A. E. Brower. 1970. Notes on the biology and distribution of moths associated with the pitcher plant in Maine. Proceedings of the Entomological Society of Ontario **101**:79–83.

Brown, J. H., J. F. Gillooly, A. P. Allen, V. M. Savage, and G. B. West. 2004. Toward a metabolic theory of ecology. Ecology **85**:1771–1789.

Bubier, J. L., T. R. Moore, and L. A. Błędzki. 2007. Effects of nutrient addition on vegetation and carbon cycling in an ombrotrophic bog. Global Change Biology **13**:1168–1186.

Buckley, H. L., J. H. Burns, J. M. Kneitel, E. L. Walters, P. Munguia, and T. E. Miller. 2004. Small-scale patterns in community structure of *Sarracenia purpurea* inquiline communities. Community Ecology **5**:181–188.

Buckley, H. L., T. E. Miller, A. M. Ellison, and N. J. Gotelli. 2003. Reverse latitudinal trends in species richness of pitcher-plant food webs. Ecology Letters **6**:825–829.

Buckley, H. L., T. E. Miller, A. M. Ellison, and N. J. Gotelli. 2010. Local to continental-scale variation in the richness and composition of an aquatic food web. Global Ecology and Biogeography **19**:711–723.

Bundy, J. G., M. P. Davey, and M. R. Viant. 2009. Environmental metabolomics: a critical review and future perspectives. Metabolomics **5**:3–21.

Burr, C. A., 1979. *The Pollination Ecology of* Sarracenia purpurea *in Cranberry Bog, Weybridge, Vermont (Addison County)*. Master's thesis, Middlebury College.

Buss, L. W. 1982. Somatic cell parasitism and the evolution of individuality. Proceedings of the National Academy of Sciences of the USA **79**:5337–5341.

Buss, L. W., and J. B. C. Jackson. 1979. Competitive networks: non-transitive competitive relationships in cryptic coral-reef environments. The American Naturalist **113**:223–234.

Butler, J. L., D. Z. Atwater, and A. M. Ellison. 2005. Red-spotted newts: an unusual nutrient source for northern pitcher plants. Northeastern Naturalist **12**:1–10.

Butler, J. L., and A. M. Ellison. 2007. Nitrogen cycling dynamics in the carnivorous northern pitcher plant, *Sarracenia purpurea*. Functional Ecology **21**:835–843.

Butler, J. L., N. J. Gotelli, and A. M. Ellison. 2008. Linking the brown and green: nutrient transformation and fate in the *Sarracenia* microecosystem. Ecology **89**:898–904.

Cain, M. L., B. G. Milligan, and A. E. Strand. 2000. Long-distance seed dispersal in plant populations. American Journal of Botany **87**:1217–1227.

Callewaert, C., S. Van Nevel, F.-M. Kerckhof, M. S. Granitsiotis, and N. Boon. 2015. Bacterial exchange in household washing machines. Frontiers in Microbiology **6**:1381.

Callicott, J. B., L. B. Crowder, and K. Mumford. 1999. Current normative concepts in conservation. Conservation Biology **13**:22–35.

Canby, W. M., 1875. *Darlingtonia californica*, an insectivorous plant. In F. W. Putnam, editor. Proceedings of the American Association for the Advancement of Science, Twenty-Third Meeting, held at Hartford Connecticut; Part II. B. Natural History, pp. 67–72. The Salem Press, Salem, MA.

Canter, E. J., C. Cuellar-Gempeler, A. I. Pastore, T. E. Miller, and O. U. Mason. 2018. Predator identity more than predator richness structures aquatic microbial assemblages in *Sarracenia purpurea* leaves. Ecology **99**:652–660.

Cardinale, B. J., J. P. Wright, M. W. Cadotte, I. T. Carroll, A. Hector, D. S. Srivastava, et al. 2007. Impacts of plant diversity on biomass production increase through time because of species complementarity. Proceedings of the National Academy of Sciences of the USA **104**:18123–18128.

Carpenter, S. R. 2003. Regime Shifts in Lake Ecosystems: Pattern and Variation. Ecology Institute, Oldendorf/Luhe, Germany.

Carpenter, S. R., and W. A. Brock. 2006. Rising variance: a leading indicator of ecological transition. Ecology Letters **9**:311–318.

Carpenter, S. R., N. F. Caraco, D. L. Correll, R. W. Howarth, A. N. Sharpley, and V. H. Smith. 1998. Nonpoint pollution of surface waters with phosphorus and nitrogen. Ecological Applications **8**:559–568.

Carpenter, S. R., J. J. Cole, J. R. Hodgson, J. F. Kitchell, M. L. Pace, D. Bade, et al. 2001. Trophic cascades, nutrients, and lake productivity: whole-lake experiments. Ecological Monographs **71**:163–186.

Carpenter, S. R., K. L. Cottingham, and D. E. Schindler. 1992. Biotic feedbacks in lake phosphorous cycles. Trends in Ecology & Evolution **7**:332–336.

Carpenter, S. R., and J. F. Kitchell. 1993. The Trophic Cascade in Lakes. Cambridge University Press, New York.

CarraDonna, P. J., and D. W. Inouye. 2015. Phenological responses to climate change do not exhibit phylogenetic signal in a subalpine plant community. Ecology **96**:355–361.

Case, F. W., and R. B. Case. 1976. The *Sarracenia rubra* complex. Rhodora **78**:270–325.

Caswell, H. 1976. Community structure: a neutral model analysis. Ecological Monographs **46**:327–354.

Caswell, H. 2006. Matrix Population Models: Construction, Analysis, and Interpretation. Sinauer Associates, Sunderland, MA.

Chamberlain, S., K. Ram, and T. Hart, 2020. spocc: interface to species occurrence data sources. https://cran.r-project.org/package=spocc.

Chandler, G. E., and J. W. Anderson. 1976. Studies on the nutrition and growth of *Drosera* species with reference to the carnivorous habit. New Phytologist **76**:129–141.

Chao, A., N. J. Gotelli, T. C. Hsieh, E. L. Sander, K. H. Ma, R. K. Colwell, and A. M. Ellison. 2014. Rarefaction and extrapolation with Hill numbers: a framework for sampling and estimation in species diversity studies. Ecological Monographs **84**:45–67.

Chapin, C. T., and J. Pastor. 1995. Nutrient limitations in the northern pitcher plant *Sarracenia purpurea*. Canadian Journal of Botany **73**:728–734.

Chapin, F. S., III, K. Autumn, and F. Pugnaire. 1993. Evolution of suites of traits in response to environmental stress. The American Naturalist **142**:S78–S92.

Chapin, F. S., III, P. M. Vitousek, and K. V. Cleve. 1986. The nature of nutrient limitation in plant communities. The American Naturalist **127**:48–58.

Charmantier, A., R. H. McCleery, L. R. Cole, C. Perrins, L. E. B. Kruuk, and B. C. Sheldon. 2008. Adaptive phenotypic plasticity in response to climate change in a wild bird population. Science **320**:800–803.

Charnov, E. L., and J. F. Gillooly. 2003. Thermal time: body size, food quality and the 10 degrees C rule. Evolutionary Ecology Research **5**:43–51.

Chase, J. M., and M. A. Leibold. 2003. Ecological Niches: Linking Classical and Contemporary Approaches. University of Chicago Press, Chicago.

Cheek, M., and M. Young. 1994. The *Limonium peregrinum* of Carolus Clusius. Carnivorous Plant Newsletter **23**:95–98.

Cicero, M. T. 2010 edition. De legibus. (Library of Latin Texts). Brepols Publishers, Turnhout, Belgium.

Cirtwill, A. R., A. Eklöf, T. Roslin, K. Wootton, and D. Gravel. 2019. A quantitative framework for investigating the reliability of empirical network construction. Methods in Ecology and Evolution **10**:902–911.

Clark, J. S. 1998. Why trees migrate so fast: confronting theory with dispersal biology and the paleorecord. The American Naturalist **152**:204–224.

Clark, J. S., D. M. Bell, M. H. Hersh, M. C. Kwit, E. Moran, C. Salk, et al. 2011. Individual-scale variation, species-scale differences: inference needed to understand diversity. Ecology Letters **14**:1273–1287.

Clark, J. S., C. Fastie, G. Hurtt, S. T. Jackson, C. Johnson, G. A. King, et al. 1998. Reid's paradox of rapid plant migration. BioScience **48**:13–24.

Clements, F. E. 1916. Plant Succession. Carnegie Institution of Washington, Washington, DC.

Clements, F. E. 2019. Nature and structure of the climax. Journal of Ecology **24**:252–284.

Clotfelter, C. T. 1976. School desegregation, "tipping," and private school enrollment. The Journal of Human Resources **11**:28–50.

Clusius, C. 1601. Rariorum plantarum historia. Plantin Press, Antwerp.

Cochran-Stafira, D. L., and C. N. von Ende. 1998. Integrating bacteria into food webs: studies with *Sarrcenia purpurea* inquilines. Ecology **79**:880–898.

Cody, M. L., and J. M. Diamond, editors. 1975. Ecology and Evolution of Communities. Harvard University Press, Cambridge, MA.

Cohen, J. E., F. Briand, and C. M. Newman. 1990. Community Food Webs: Data and Theory. Springer-Verlag, Berlin, Germany.

Cohen, J. E. 2010. Ecologists' Co-Operative Web Bank, Version 1.1. The Rockefeller University, New York.

Collier, M. J., and C. Devitt. 2016. Novel ecosystems: challenges and opportunities for the Anthropocene. Anthropocene Review **3**:231–242.

Colwell, R. K., G. Brehm, C. L. Cardelús, A. C. Gilman, and J. T. Longino. 2008. Global warming, elevational range shifts, and lowland biotic attrition in the wet tropics. Science **322**:258–261.

Connell, J. H., and R. O. Slatyer. 1977. Mechanisms of succession in natural communities and their role in community stability and organization. The American Naturalist **111**:1119–1144.

Connor, E. F., and D. Simberloff. 1979. The assembly of species communities: chance or competition? Ecology **60**:1132–1140.

Contamin, R., and A. M. Ellison. 2009. Indicators of regime shifts in ecological systems: what do we need to know and when do we need to know it? Ecological Applications **19**:799–816.

Cooper, G. S., S. Willcock, and J. A. Dearing. 2020. Regime shifts occur disproportionately faster in larger ecosystems. Nature Communications **11**:1175.

Cottenie, K. 2005. Integrating environmental and spatial processes in ecological community dynamics. Ecology Letters **8**:1175–1182.

Cottingham, K. L., J. T. Lennon, and B. L. Brown. 2005. Knowing when to draw the line: designing more informative ecological experiments. Frontiers in Ecology and the Environment **3**:145–152.

Couwenberg, J., and H. Joosten. 2004. News from Ireland: alien invader on the Bog of Allen. International Mire Conservation Group Newsletter **2004**(4):27–28.

Cowles, H. C. 1899. The Ecological Relations of the Vegetation on the Sand Dunes of Lake Michigan. University of Chicago Press, Chicago.

Cowles, H. C. 1911. The causes of vegetation cycles. Annals of the Association of American Geographers **1**:3–20.

Cowper, W. 1785. The Task: A Poem. J. Johnson, London.

Cox, J., M. Ellershaw, J. Erian, E. Goldberg, J. Hosking, A. Jowitt, et al. 2015. Invasive Species Theme Plan: Strategic Principles for the Management of Invasive Species on Natura 2000 Sites. Natural England, London.

Cresswell, J. E. 1991. Capture rates and composition of insect prey of the pitcher plant *Sarracenia purpurea*. The American Midland Naturalist **125**:1–9.

Cresswell, J. E. 1993. The morphological correlates of prey capture and resource parasitism in pitchers of the carnivorous plant *Sarracenia purpurea*. The American Midland Naturalist **129**:35–41.

Croizat, L. 1952. Manual of Phytogeography, or, An Account of Plant-Dispersal throughout the World. Dr. W. Junk, The Hague.

Cross, A. T., A. R. Davis, A. Fleischmann, J. D. Horner, A. Jürgens, D. J. Merritt, et al. 2018. Evolution of carnivory in angiosperms. In A. M. Ellison and L. Adamec, editors. Carnivorous Plants: Physiology, Ecology, and Evolution, pp. 294–313. Oxford University Press, Oxford.

Crouse, D., L. B. Crowder, and H. Caswell. 1987. A stage-based population model for loggerhead sea turtles and implications for conservation. Ecology **68**:1412–1423.

Currie, D. J., and S. Venne. 2016. Climate change is not a major driver of shifts in the geographical distributions of North American birds. Global Ecology and Biogeography **26**:333–346.

Dahlem, G. A., and R. F. C. Naczi. 2006. Flesh flies (Diptera: Sarcophagidae) associated with North American pitcher plants (Sarraceniaceae), with descriptions of three new species. Annals of the Entomological Society of America **99**:218–240.

Dai, L., K. S. Korolev, and J. Gore. 2013. Slower recovery in space before collapse of connected populations. Nature **496**:355–358.

Dakos, V., S. R. Carpenter, W. A. Brock, A. M. Ellison, V. Guttal, A. R. Ives, et al. 2012. Methods for detecting early warnings of critical transitions in time series illustrated using simulated ecological data. PLoS ONE **7**:e41010.

Dakos, V., B. Matthews, A. P. Hendry, J. Levine, N. Loeuille, J. Norberg, et al. 2019. Ecosystem tipping points in an evolving world. Nature Ecology and Evolution **3**:355–362.

Dakos, V., M. Scheffer, E. H. van Nes, V. Brovkin, V. Petoukhov, and H. Held. 2008. Slowing down as an early warning signal for abrupt climate change. Proceedings of the National Academy of Sciences of the USA **105**:14308–14312.

Dang, H. Y., and C. R. Lovell. 2016. Microbial surface colonization and biofilm development in marine environments. Microbiology and Molecular Biology Reviews **80**:91–138.

Darwin, C. 1875. Insectivorous Plants. John Murray, London.

Datry, T., N. Bonada, and J. Heino. 2016. Towards understanding the organisation of metacommunities in highly dynamic ecological systems. Oikos **125**:149–159.

Davis, C. C., C. G. Willis, B. Connolly, C. Kelly, and A. M. Ellison. 2015. Herbarium records are reliable sources of phenological change driven by climate and provide novel insights into species' phenological cueing mechanisms. American Journal of Botany **102**:1599–1609.

Davis, S. J., and K. Caldeira. 2010. Consumption-based accounting of CO_2 emissions. Proceedings of the National Academy of Sciences of the USA **107**:5687–5692.

de Groot, R. S., M. A. Wilson, and R. M. J. Boumans. 2002. A typology for the classification, description and valuation of ecosystem functions, goods and services. Ecological Economics **41**:393–408.

de Souza, A. C., R. C. Q. Portela, and E. A. de Mattos. 2018. Demographic processes limit upward altitudinal range expansion in a threatened tropical palm. Ecology and Evolution **8**:12238–12249.

Del Moral, R., and L. C. Bliss. 1993. Mechanisms of primary succession: insights resulting from the eruption of Mount St. Helens. Advances in Ecological Research **24**:1–66.

DeNiro, M. J., and S. Epstein. 1976. You are what you eat (plus a few ‰): the carbon isotope cycle in food chains. Geological Society of America, Abstracts with Programs **1976**:834–835.

Dézerald, O., C. Leroy, B. Corbara, J.-F. Carrias, L. Pélozuelo, A. Dejean, and R. Céréghino. 2013. Food-web structure in relation to environmental gradients and predator-prey ratios in tank-bromeliad ecosystems. PLoS ONE **8**:e71735.

Diamond, J. 1986. Overview: laboratory experiments, field experiments, and natural experiments. In T. J. Case and J. Diamond, editors. Community Ecology, pp. 3–22. Harper & Row, New York.

Diamond, J. M. 1975. Assembly of species communities. In M. L. Cody and J. M. Diamond, editors. Ecology and Evolution of Communities, pp. 342–444. Harvard University Press, Cambridge, MA.

Diamond, S. E., L. M. Nichols, S. L. Pelini, C. A. Penick, G. W. Barber, S. H. Cahan, et al. 2016. Climatic warming destabilizes forest ant communities. Science Advances **2**:e1600842.

Díaz, S., J. Kattge, J. H. C. Cornelissen, I. J. Wright, S. Lavorel, S. Dray, et al. 2016. The global spectrum of plant form and function. Nature **529**:167–171.

Dixon, P. M., A. M. Ellison, and N. J. Gotelli. 2005. Improving the precision of estimates of the frequency of rare events. Ecology **86**:1114–1123.

Doak, D. F., K. Gross, and W. F. Morris. 2005. Understanding and predicting the effects of sparse data on demographic analyses. Ecology **86**:1154–1163.

Doak, D. F., and W. Morris. 1999. Detecting population-level consequences of ongoing environmental change without long-term monitoring. Ecology **80**:1537–1551.

Doncaster, C. P., V. A. Chávez, C. Viguier, R. Wang, E. Zhang, X. Dong, et al. 2016. Early warning of critical transitions in biodiversity from compositional disorder. Ecology **97**:3079–3090.

Donohue, I., H. Hillebrand, J. M. Montoya, O. L. Petchey, S. L. Pimm, M. S. Fowler, et al. 2016. Navigating the complexity of ecological stability. Ecology Letters **19**:1172–1185.

Dornelas, M., and The BioTIME. Consortium. 2018. BioTIME: a database of biodiversity time series for the Anthropocene. Global Ecology and Biogeography **27**:760–786.

Dornelas, M., N. J. Gotelli, H. Shimadzu, F. Moyes, A. E. Magurran, and B. J. McGill. 2019. A balance of winners and losers in the Anthropocene. Ecology Letters **22**:847–854.

Dornelas, M., A. E. Magurran, S. T. Buckland, A. Chao, R. L. Chazdon, R. K. Colwell, et al. 2013. Quantifying temporal change in biodiversity: challenges and opportunities. Proceedings of the Royal Society B: Biological Sciences **280**:20121931.

Drake, J., T. E. Flum, G. J. Wittemann, T. Voskuil, A. M. Hoylman, C. Creson, et al. 1993. The construction and assembly of an ecological landscape. Journal of Animal Ecology **62**:117–130.

Drake, J. A. 1990. Communities as assembled structures: do rules govern patterns? Trends in Ecology & Evolution **5**:159–164.

Drake, J. M., and B. D. Griffen. 2010. Early warning signals of extinction in deteriorating environments. Nature **467**:456–459.

Drury, W., and I. Nisbet. 1973. Succession. Journal of the Arnold Arboretum **54**:331–368.

Dudgeon, S., and P. S. Petraitis. 2005. First year demography of the foundation species, *Ascophyllum nodosum*, and its community implications. Oikos **109**:405–415.

Dullinger, S., A. Gattringer, W. Thuiller, D. Moser, N. E. Zimmermann, A. Guisan, et al. 2012. Extinction debt of high-mountain plants under twenty-first-century climate change. Nature Climate Change **2**:619–622.

Dunne, J. A., and R. J. Williams. 2009. Cascading extinctions and community collapse in model food webs. Philosophical Transactions of the Royal Society, B: Biological Sciences **364**:1711–1723.

Dunne, J. A., R. J. Williams, and N. D. Martinez. 2002. Food-web structure and network theory: the role of connectance and size. Proceedings of the National Academy of Sciences of the USA **99**:12917–12922.

Ebenman, B., and T. Jonsson. 2005. Using community viability analysis to identify fragile systems and keystone species. Trends in Ecology & Evolution **20**:568–575.

Eby, S., A. Agrawal, S. Majumder, A. P. Dobson, and V. Guttal. 2017. Alternative stable states and spatial indicators of critical slowing down along a spatial gradient in a savanna ecosystem. Global Ecology and Biogeography **28**:636–649.

Efron, B. 2005. Bayesians, frequentists, and scientists. Journal of the American Statistical Association **100**:1–5.

Egerton, F. N. 2012. Roots of Ecology: Antiquity to Haeckel. University of California Press, Berkeley.

Egler, F. E. 1954. Vegetation science concepts: I. Initial floristic composition, a factor in old field vegetation development. Vegetatio **4**:412–417.

Ehrlén, J., and W. Morris. 2015. Predicting changes in the distribution and abundance of species under environmental change. Ecology Letters **18**:303–314.

Elith, J., and J. R. Leathwick. 2009. Species distribution models: ecological explanation and prediction across space and time. Annual Review of Ecology, Evolution, and Systematics **40**:677–697.

Ellison, A. M. 1996. An introduction to Bayesian inference for ecological research and environmental decision-making. Ecological Applications **6**:1036–1046.

Ellison, A. M. 2001. Interpsecific and intraspecific variation in seed size and germination requirements of *Sarracenia* (Sarraceniaceae). American Journal of Botany **88**:429–437.

Ellison, A. M. 2006. Nutrient limitation and stoichiometry of carnivorous plants. Plant Biology **8**:740–747.

Ellison, A. M. 2013. The suffocating embrace of landscape and the picturesque conditioning of ecology. Landscape Journal **32**:79–94.

Ellison, A. M. 2018. A sense of scale. Bulletin of the Ecological Society of America **99**:173–179.

Ellison, A. M. 2019. Foundation species, non-trophic interactions, and the value of being common. iScience **13**:254–268.

Ellison, A. M. 2020. Foraging modes of carnivorous plants. Israel Journal of Ecology and Evolution **66**:101–112.

Ellison, A. M., and L. Adamec. 2011. Ecophysiological traits of terrestrial and aquatic carnivorous plants: are the costs and benefits the same? Oikos **120**:1721–731.

Ellison, A. M., and L. Adamec. 2018. Introduction: what is a carnivorous plant? In A. M. Ellison and L. Adamec, editors. Carnivorous Plants: Physiology, Ecology, and Evolution, pp. 3–6. Oxford University Press, Oxford.

Ellison, A. M., H. L. Buckley, T. E. Miller, and N. J. Gotelli. 2004. Morphological variation in *Sarracenia purpurea* (Sarraceniaceae): geographic, environmental, and taxonomic correlates. American Journal of Botany **91**:1930–1935.

Ellison, A. M., and L. W. Buss. 1983. A naturally occurring developmental synergism between the cellular slime mold, *Dictyostelium mucoroides* and the fungus, *Mucor hiemalis*. American Journal of Botany **70**:298–302.

Ellison, A. M., E. D. Butler, E. J. Hicks, R. F. C. Naczi, P. J. Calie, C. D. Bell, and C. C. Davis. 2012. Phylogeny and biogeography of the carnivorous plant family Sarraceniaceae. PLoS ONE **7**:e39291.

Ellison, A. M., C. C. Davis, P. J. Calie, and R. F. C. Naczi. 2014. Pitcher plants (*Sarracenia*) provide a 21st-century perspective on infraspecific ranks and interspecific hybrids: a modest proposal for appropriate recognition and usage. Systematic Botany **93**:939–949.

Ellison, A. M., P. M. Dixon, and J. Ngai. 1994. A null model for neighborhood models of plant competitive interactions. Oikos **71**:225–238.

Ellison, A. M., and E. J. Farnsworth, 2001. Mangrove communities. In M. D. Bertness, S. Gaines, and M. E. Hay, editors. Marine Community Ecology, pp. 423–442. Sinauer Associates, Sunderland, MA.

Ellison, A. M., and E. J. Farnsworth. 2005. The cost of carnivory for *Darlingtonia californica* (Sarraceniaceae): evidence from relationships among leaf traits. American Journal of Botany **92**:1085–1093.

Ellison, A. M., and N. J. Gotelli. 2001. Evolutionary ecology of carnivorous plants. Trends in Ecology & Evolution **16**:623–629.

Ellison, A. M., and N. J. Gotelli. 2002. Nitrogen availability alters the expression of carnivory in the northern pitcher plant *Sarracenia purpurea*. Proceedings of the National Academy of Sciences of the USA **99**:4409–4412.

Ellison, A. M., and N. J. Gotelli. 2009. Energetics and the evolution of carnivorous plants: Darwin's "most wonderful plants in the world." Journal of Experimental Botany **60**:19–42.

Ellison, A. M., N. J. Gotelli, L. A. Błędzki, and J. L. Butler. 2021. Regulation by the pitcher plant *Sarracenia purpurea* of the structure of its inquiline food web. American Midland Naturalist doi: 10.1101/2020.05.15.098004.

Ellison, A. M., N. J. Gotelli, J. S. Brewer, L. Cochran-Stafira, J. Kneitel, T. E. Miller, A. S. Worley, and R. Zamora. 2003. The evolutionary ecology of carnivorous plants. Advances in Ecological Research **33**:1–74.

Ellison, A. M., and J. N. Parker. 2002. Seed dispersal and seedling establishment of *Sarracenia purpurea* (Sarraceniaceae). American Journal of Botany **89**: 1024–1026.

Ellner, S. P., and M. Rees. 2006. Integral projection models for species with complex demography. The American Naturalist **167**:410–428.

Elser, J. J., W. F. Fagan, A. J. Kerkhoff, N. G. Swenson, and B. J. Enquist. 2010. Biological stoichiometry of plant production: metabolism, scaling and ecological response to global change. New Phytologist **186**:593–608.

Elton, C. S. 1927. Animal Ecology. Sidgwick and Jackson, London.

Elton, C. S. 1958. The Ecology of Invasions by Animals and Plants. Chapman & Hall, London.

Endler, J. A. 1986. Natural Selection in the Wild. Princeton University Press, Princeton, NJ.

Enquist, B. J., E. P. Economo, T. E. Huxman, A. P. Allen, D. D. Ignace, and J. F. Gillooly. 2003. Scaling metabolism from organisms to ecosystems. Nature **423**:639–642.

Erdös, P., and A. Rényi. 1959. On random graphs. I. Publicationes Mathematicae **6**:290–297.

Estes, L., P. R. Elsen, T. Treur, L. Ahmed, K. Caylor, J. Chang, J. J. Choi, and E. C. Ellis. 2018. The spatial and temporal domains of modern ecology. Nature Ecology and Evolution **2**:819–826.

Evans, C., and T. D. Davies. 1998. Causes of concentration/discharge hysteresis and its potential as a tool for analysis of episode hydrochemistry. Water Resources Research **34**:129–137.

Evers, C. R., C. B. Wardropper, B. Branoff, E. F. Granek, S. L. Hirsch, T. E. Link, et al. 2018. The ecosystem services and biodiversity of novel ecosystems: a literature review. Global Ecology and Conservation **13**:e00362.

Faber-Langendoen, D., K. Baldwin, R. K. Peet, D. Meidinger, E. Muldavin, T. Keeler-Wolf, and C. Josse. 2018. The EcoVeg approach in the Americas: US, Canadian and international vegetation classifications. Phytocoenologia **48**:215–237.

Fajardo, A., and A. Siefert. 2018. Intraspecific trait variation and the leaf economics spectrum across resource gradients and levels of organization. Ecology **99**:1024–1030.

Farjalla, V. F., A. L. González, R. Cereghino, O. Dezerald, N. A. C. Marino, G. C. O. Piccoli, et al. 2016. Terrestrial support of aquatic food webs depends on light inputs: a geographically-replicated test using tank bromeliads. Ecology **97**:2147–2156.

Farnsworth, E. J., M. Chu, W. J. Kress, A. K. Neill, J. H. Best, R. D. Stevenson, et al. 2013. Next-generation field guides. BioScience **63**:891–899.

Farnsworth, E. J., and A. M. Ellison. 2008. Prey availability directly affects physiology, growth, nutrient allocation, and scaling relationships among leaf traits in ten carnivorous plant species. Journal of Ecology **96**:213–221.

Farnsworth, K. D., L. Albantakis, and T. Caruso. 2017. Unifying concepts of biological function from molecules to ecosystems. Oikos **126**:1367–1376.

Farrell, T. M. 1991. Models and mechanisms of succession: an example from a rocky intertidal community. Ecological Monographs **61**:95–113.

Fashing, N. J. 1994. A new species of *Leipothrix* (Prostigmata: Eriophyidae) from the cobra lily *Darlingtonia californica* (Sarraceniaceae). International Journal of Acarology **20**:99–101.

Fashing, N. J., and B. M. OConnor. 1984. *Sarraceniopus*—a new genus for histiostomatid mites inhabiting the pitchers of the Sarraceniaceae (Astigmata: Histiostomatidae). International Journal of Acarology **10**:217–227.

Fastie, C. L. 1995. Causes and ecosystem consequences of multiple pathways of primary succession at Glacier Bay, Alaska. Ecology **76**:1899–1916.

Feldmeyer, E. 1985. Étude phyto-écologique de la tourbi ère des Tenasses. Botanica Helvetica **95**:99–115.

Fernandes, I. M., R. Henriques-Silva, J. Penha, J. Zuanon, and P. R. Peres-Neto. 2013. Spatiotemporal dynamics in a seasonal metacommunity structure is predictable: the case of floodplain-fish communities. Ecography **37**:464–475.

Fersht, A. R. 2017. Structure and Mechanism In Protein Science: A Guide To Enzyme Catalysis And Protein Folding. World Scientific, Singapore.

Fick, S. E., and R. J. Hijmans. 2017. Worldclim 2: new 1-km spatial resolution climate surfaces for global land areas. International Journal of Climatology **37**:4302–4315.

Fields, S., and M. Johnston. 2005. Whither model organism research? Science **307**:1885–1886.

Filewod, B., and S. C. Thomas. 2014. Impacts of a spring heat wave on canopy processes in a northern hardwood forest. Global Change Biology **20**:360–371.

Fish, D. 1976. Insect-plant relationships of the insectivorous pitcher plant *Sarracenia minor*. The Florida Entomologist **59**:199–203.

Fish, D., and D. W. Hall. 1978. Succession and stratification of aquatic insects inhabiting the leaves of the insectivorous pitcher plant *Sarracenia purpurea*. The American Midland Naturalist **99**:172–183.

Fitzpatrick, M. C., and A. M. Ellison, 2018. Estimating the exposure of carnivorous plants to rapid climatic change. In A. M. Ellison and L. Adamec, editors. Carnivorous Plants: Physiology, Ecology, and Evolution, pp. 389–407. Oxford University Press, Oxford.

Fitzpatrick, M. C., N. J. Gotelli, and A. M. Ellison. 2013. MaxEnt vs. MaxLike: empirical comparisons with ant species distributions. Ecosphere **4**:55.

Fitzpatrick, M. C., and S. R. Keller. 2015. Ecological genomics meets community-level modelling of biodiversity: mapping the genomic landscape of current and future environmental adaptation. Ecology Letters **18**:1–16.

Fleischmann, A., J. Schlauer, S. A. Smith, and T. J. Givnish, 2018. Evolution of carnivory in angiosperms. In A. M. Ellison and L. Adamec, editors. Carnivorous Plants: Physiology, Ecology, and Evolution, pp. 22–42. Oxford University Press, Oxford.

Fogarty, M. J., R. Gamble, and C. T. Perretti. 2016. Dynamic complexity in exploited marine ecosystems. Frontiers in Ecology and Evolution **4**:68.

Fois, M., A. Cuena-Lombrana, G. Fenu, D. Cogoni, and G. Bacchetta. 2018. Does a correlation exist between environmental suitability models and plant population parameters? An experimental approach to measure the influence of disturbances and environmental changes. Ecological Indicators **86**:1–8.

Folkerts, D. 1999. Pitcher plant wetlands of the southeastern United States. Arthropod associates. In D. P. Batzer, R. B. Rader, and S. A. Wissinger, editors. Invertebrates in Freshwater Wetlands of North America: Ecology and Management, pp. 247–275. John Wiley & Sons, New York.

Folkerts, D. R. 1992. *Interactions of Pitcher Plants* (Sarracenia: *Sarraceniaceae*) *with their Arthropod Prey in the Southeastern United States*. PhD dissertation, University of Georgia.

Folkerts, D. R., and G. W. Folkerts. 1996. Aids for field identification of pitcher plant moths of the genus *Exyra* (Lepidoptera: Noctuidae). Entomological News **107**:128–136.

Forbes, S. A. 1887. The lake as a microcosm. Bulletin of the Scientific Association of Peoria, Illinois **1887**:77–87.

Foss, P. J., and C. A. O'Connell. 1985. Notes on the ecology of *Sarracenia purpurea* L. on Irish peatlands. The Irish Naturalists' Journal **21**:440–443.

Foster, D. R., W. W. Oswald, E. K. Faison, E. D. Doughty, and B. C. S. Hansen. 2006. A climatic driver for abrupt mid-Holocene vegetation dynamics and the hemlock decline in New England. Ecology **87**:2959–2966.

Foster, G., D. Royer, and D. Lunt. 2017. Future climate forcing potentially without precedent in the last 420 million years. Nature Communications **8**:14845.

Franck, D. H. 1976. The morphological interpretation of epiascidiate leaves. Botanical Review **42**:345–388.

Franco, M., and J. Silvertown. 2004. Comparative demography of plants based upon elasticities of vital rates. Ecology **85**:531–538.

Frank, J. H., and L. P. Lounibos. 1983. Phytotelmata: Terrestrial Plants as Hosts for Aquatic Insect Communities. Plexus Press, Medford, MA.

Frank, K. T., B. Petrie, and N. L. Shackell. 2007. The ups and downs of trophic control in continental shelf ecosystems. Trends in Ecology & Evolution **22**:236–242.

Franks, S., S. Sim, and A. Weis. 2007. Rapid evolution of flowering time by an annual plant in response to a climate fluctuation. Proceedings of the National Academy of Sciences of the USA **104**:1278–1282.

Freeman, J., L. Kobziar, E. W. Rose, and W. Cropper. 2017. A critique of the historical-fire-regime concept in conservation. Conservation Biology **31**:976–985.

Gallie, D. R., and S.-C. Chang. 1997. Signal transduction in the carnivorous plant *Sarracenia purpurea*. Plant Physiology **115**:1461–1471.

Gardner, M. R., and W. R. Ashby. 1970. Connectance of large dynamic (cybernetic) systems: critical values for stability. Nature **228**:784.

Gibson, T. C., 1983. *Competition, Disturbance, and the Carnivorous Plant Community in the Southeastern United States*. PhD dissertation, The University of Utah.

Gilpin, M. E. 1975. Stability of feasible predator-prey systems. Nature **254**:137–139.

Gilpin, M. E., and J. M. Diamond. 1982. Factors contributing to non-randomness in species co-occurrences on islands. Oecologia **52**:75–84.

Givnish, T., 1979. On the adaptive significance of leaf form. In O. Solbrig, S. Jain, G. Johnson, and H. R. P, editors. Topics in Plant Population Biology pp. 375–407. Columbia University Press, New York.

Givnish, T. J., editor. 1986. On the Economy of Plant Form and Function. Cambridge University Press, Cambridge.

Givnish, T. J. 1989. Ecology and evolution of carnivorous plants. In W. G. Abrahamson, editor. Plant-Animal Interactions, pp. 343–290. McGraw-Hill, New York.

Givnish, T. J., E. L. Burkhardt, R. E. Happel, and J. D. Weintraub. 1984. Carnivory in the bromeliad *Brocchinia reducta*, with a cost/benefit model for the general restriction of carnivorous plants to sunny, moist, nutrient-poor habitats. The American Naturalist **124**:479–497.

Givnish, T. J., K. W. Sparks, S. J. Hunter, and A. Pavlovič, 2018. Why are plants carnivorous? Cost/benefit analysis, whole plant growth, and the context-specific advantages of botanical carnivory. In A. M. Ellison and L. Adamec, editors. Carnivorous Plants: Physiology, Ecology, and Evolution, pp. 232–255. Oxford University Press, Oxford.

Givnish, T. J., and G. J. Vermeij. 1976. Sizes and shapes of liane leaves. The American Naturalist **110**:742–778.

Gladwell, M. 2000. The Tipping Point: How Little Things Can Make a Big Difference. Little Brown & Company, New York.

Gleason, H. A. 1926. The individualistic concept of the plant association. Bulletin of the Torrey Botanical Club **53**:7–26.

Gleason, H. A., and A. Cronquist. 1991. Manual of Vascular Plants of Northeastern United States and Adjacent Canada. New York Botanical Garden, Bronx, New York.

Godt, M. J. W., and J. L. Hamrick. 1996. Genetic structure of two endangered pitcher plants, *Sarracenia jonesii* and *Sarracenia oreophila* (Sarraceniaceae). American Journal of Botany **83**:1016–1023.

Godt, M. J. W., and J. L. Hamrick. 1998a. Allozyme diversity in the endangered pitcher plants *Sarracenia rubra* ssp. *alabamensis* (Sarraceniaceae) and its close relative *Sarracenia rubra* ssp. *rubra*. American Journal of Botany **85**:802–810.

Godt, M. J. W., and J. L. Hamrick. 1998b. Genetic divergence among infraspecific taxa of *Sarracenia purpurea*. Systematic Botany **23**:427–438.

Goebel, K., 1891. Insektivoren. In K. Goebel, editor. Pflanzenbiologische Schilderungen, pp. 53–214. N. G. Elwert'sche, Marburg, Germany.

Goetze, J. S., J. Claudet, F. Januchowski-Hartley, T. J. Langlois, S. K. Wilson, C. White, et al. 2018. Demonstrating multiple benefits from periodically harvested fisheries closures. Journal of Applied Ecology **55**:1102–1113.

González, A. L., O. Dézerald, P. A. Marquet, G. Q. Romero, and D. S. Srivastava. 2017. The multidimensional stoichiometric niche. Frontiers in Ecology and Evolution **5**:110.

González, A. L., J. M. Fariña, R. Pinto, C. Pérez, K. C. Weathers, J. J. Armesto, and P. A. Marquet. 2011. Bromeliad growth and stoichiometry: responses to atmospheric nutrient supply in fog-dependent ecosystems of the hyperarid Atacama Desert, Chile. Oecologia **167**:835–845.

González, A. L., G. Q. Romero, and D. S. Srivastava. 2014. Detrital nutrient content determines growth rate and elemental composition of bromeliad-dwelling insects. Freshwater Biology **59**:737–747.

González, F., and N. Pabón-Mora. 2015. Trickery flowers: the extraordinary chemical mimicry of *Aristolochia* to accomplish deception to its pollinators. New Phytologist **206**:10–13.

Gordon, D. M. 2010. Ant Encounters: Interaction Networks and Colony Behavior. Princeton University Press, Princeton, NJ.

Gore, A. 1968. The supply of six elements by rain to an uplant peat area. Journal of Ecology **56**:583–495.

Gotelli, N. J. 1997. Competition and coexistence of larval ant lions. Ecology **78**:1761–1773.

Gotelli, N. J. 1999. How do communities come together? Science **286**:1684–1685.

Gotelli, N. J. 2000. Null model analysis of species co-occurrence patterns. Ecology **81**:2606–2621.

Gotelli, N. J., and R. K. Colwell. 2001. Quantifying biodiversity: procedures and pitfalls in the measurement and comparison of species richness. Ecology Letters **4**:379–391.

Gotelli, N. J., and A. M. Ellison. 2002. Nitrogen deposition and extinction risk in the northern pitcher plant *Sarracenia purpurea*. Ecology **83**:2758–2765.

Gotelli, N. J., and A. M. Ellison. 2006a. Food-web models predict species abundance in response to habitat change. PLoS Biology **44**:e324.

Gotelli, N. J., and A. M. Ellison. 2006b. Forecasting extinction risk with non-stationary matrix models. Ecological Applications **16**:51–61.

Gotelli, N. J., A. M. Ellison, and B. A. Ballif. 2012. Environmental proteomics, biodiversity statistics and food-web structure. Trends in Ecology & Evolution **27**:436–442.

Gotelli, N. J., and G. R. Graves. 1996. Null Models in Ecology. Smithsonian Institution Press, Washington, DC.

Gotelli, N. J., and D. J. McCabe. 2002. Species co-occurrence: a meta-analysis of J. M. Diamond's assembly rules model. Ecology **83**:2091–2096.

Gotelli, N. J., and B. J. McGill. 2006. Null versus neutral models: what's the difference? Ecography **29**:793–800.

Gotelli, N. J., P. J. Mouser, S. P. Hudman, S. E. Morales, D. S. Ross, and A. M. Ellison. 2008. Geographic variation in nutrient availability, stoichiometry, and metal concentrations of plants and pore-water in ombrotrophic bogs in New England, USA. Wetlands **28**:827–840.

Gotelli, N. J., H. Shimadzu, M. Dornelas, B. McGill, F. Moyes, and A. E. Magurran. 2017. Community-level regulation of temporal trends in biodiversity. Science Advances **3**:e1700315.

Gotelli, N. J., A. M. Smith, A. M. Ellison, and B. A. Ballif. 2011. Proteomic characterization of the major arthropod associates of the carnivorous pitcher plant *Sarracenia purpurea*. Proteomics **11**:2354–2358.

Gotelli, N. J., and J. Stanton-Geddes. 2015. Climate change, genetic markers and species distribution modelling. Journal of Biogeography **42**:1577–1585.

Gotelli, N. J., and W. Ulrich. 2010. The empirical Bayes approach as a tool to identify non-random species associations. Oecologia **162**:463–477.

Gotelli, N. J., and W. Ulrich. 2012. Statistical challenges in null model analysis. Oikos **121**:171–180.

Gotsch, S. G., and A. M. Ellison. 1998. Seed germination of the northern pitcher plant, *Sarracenia purpurea*. Northeastern Naturalist **5**:175–182.

Graham, C., S. Ferrier, F. Huettman, C. Moritz, and A. Peterson. 2004. New developments in museum-based informatics and applications in biodiversity analysis. Trends in Ecology & Evolution **19**:497–503.

Grant, P. R., and I. Abbott. 1980. Interspecific competition, island biogeography and null hypotheses. Evolution **34**:332–341.

Gray, A. 1876. Darwiniana. Appleton, New York.

Gray, S., D. Akob, S. Green, and J. Kostka. 2012. The bacterial composition within the *Sarracenia purpurea* model system: local scale differences and the relationship with the other members of the food web. PLoS ONE **7**:e50969.

Gray, S. M. 2012. Succession in the aquatic *Sarracenia purpurea* community: deterministic or driven by contingency? Aquatic Ecology **46**:487–499.

Gray, S. M., T. Poisot, E. Harvey, N. Moquet, T. E. Miller, and D. Gravel. 2016. Temperature and trophic structure are driving microbial productivity along a biogeographic gradient. Ecography **39**:981–989.

Greenbaum, D. 2018. The Clean Air Act: substantial success and the challenges ahead. Annals of the American Thoracic Society **15**:296–297.

Griffith, C. B. 1960. The Passionate People Eater. [Screenplay for the movie The Little Shop of Horrors]. The Filmgroup/Santa Clara Productions, Santa Clara.

Grigg, G. C., and F. Seebacher. 1999. Field test of a paradigm: hysteresis of heart rate in thermoregulation by a free-ranging lizard (*Pogona barbata*). Proceedings of the Royal Society B: Biological Sciences **266**:1291–1297.

Grilli, J., M. Adorisio, S. Suweis, G. Barabas, J. R. Banavar, S. Allesina, and A. Maritan. 2017. Feasibility and coexistence of large ecological communities. Nature Communications **8**:16228.

Grimm, V., and C. Wissell. 1997. Babel, or the ecological stability discussions: an inventory and analysis of terminology and a guide for avoiding confusion. Oecologia **109**:323–334.

Grodzins, M. 1957. Metropolitan segregation. Scientific American **197**:33–41.

Gronberg, J. A. M., A. S. Ludtke, and D. L. Knifong, 2014. Estimates of Inorganic Nitrogen Wet Deposition from Precipitation for the Coterminous United States, 1955-1984. Technical Report 2014-5067, US Geological Survey. http://dx.doi.org/10.3133/sir20145067.

Grothjan, J. J., and E. B. Young. 2019. Diverse microbial communities hosted by the model carnivorous pitcher plant *Sarracenia purpurea*: analysis of both bacterial and eukaryotic composition across distinct host plant populations. PeerJ **7**:e6392.

Gruber, N., and S. C. Doney, 2009. Modeling of ocean biogeochemistry and ecology. In K. K. T. J. Steele and S. A. Thorpe, editors. Encyclopedia of Ocean Sciences, 2nd ed., pp. 89–104. Elsevier B. V., Amsterdam.

Gruner, D. S., J. E. Smith, E. W. Seabloom, S. A. Sandin, J. T. Ngai, H. Hillebrand, et al. 2008. A cross-system synthesis of consumer and nutrient resource control on producer biomass. Ecology Letters **11**:740–755.

Guercio, R. A., A. Shevchenko, A. Shevchenko, J. L. López-Lozano, J. Paba, M. V. Sousa, and C. A. Ricart. 2006. Ontogenetic variations in the venom proteome of the Amazonian snake *Bothrops atrox*. Proteome Science **4**:11.

Guisan, A., W. Thuiller, and N. E. Zimmermann. 2017. Habitat Suitability and Distribution Models, with Applications in R. Cambridge University Press, Cambridge.

Guo, J., and C. T. Halson. 2020. Stigma, pollen tube transmitting tract, and epidermal micromorphology of the style of *Sarracenia purpurea* (Sarraceniaceae). Botany **98**:209–229.

Gusewell, S. 2004. N:P ratios in terrestrial plants: variation and functional significance. New Phytologist **164**:243–266.

Guttal, V., C. Jayaprakash, and O. P. Tabbaa. 2013. Robustness of early warning signals of regime shifts in time-delayed ecological models. Theoretical Ecology **6**:271–283.

Haller, L., M. Tonolla, J. Zopfi, R. Peduzzi, W. Wildi, and J. Pote. 2011. Composition of bacterial and archaeal communities in freshwater sediments with different contamination levels (Lake Geneva, Switzerland). Water Research **45**:1213–1228.

Hampe, A., and R. J. Petit. 2005. Conserving biodiversity under climate change: the rear edge matters. Ecology Letters **8**:461–467.

Hardy, O. J. 2008. Testing the spatial phylogenetic structure of local communities: statistical performances of different null models and test statistics on a locally neutral community. Journal of Ecology **96**:914–926.

Harper, J. 1977. Population Biology of Plants. Academic Press, London.

Harris, D. J. 2016. Inferring species interactions from co-occurrence data with Markov networks. Ecology **97**:3308–3314.

Harvey, E., I. Gounand, E. A. Fronhofer, and A. Altermatt. 2020. Metaecosystem dynamics drive community composition in experimental, multi-layered spatial networks. Oikos **129**:402–412.

Harvey, E., and T. E. Miller. 1996. Variance in composition of inquiline communities in leaves of *Sarracenia purpurea* L. on multiple spatial scales. Oecologia **108**:562–566.

Harvey, J. W., P. R. Wetzel, T. E. Lodge, V. C. Engel, and M. S. Ross. 2017. Role of a naturally varying flow regime in Everglades restoration. Restoration Ecology **25**:S27–S38.

Hastings, A., K. C. Abbott, K. Cuddington, T. Francis, G. Gellner, Y.-C. Lai, et al. 2018. Transient phenomena in ecology. Science **361**:eaat6412.

Hastings, A., and D. B. Wysham. 2010. Regime shifts in ecological systems can occur without warning. Ecology Letters **13**:464–472.

Hawkins, B. A., R. Field, H. V. Cornell, D. J. Currie, J. F. Guegan, D. M. et al. 2003. Energy, water, and broad-scale geographic patterns of species richness. Ecology **84**:3105–3117.

Heard, S. B. 1994. Pitcher-plant midges and mosquitoes: a processing chain commensalism. Ecology **139**:79–89.

Heard, S. B. 1998. Capture rates of invertebrate prey by the pitcher plant *Sarracenia purpurea* L. The American Midland Naturalist **139**:79–89.

Heenan, P. B., P. J. de Lange, E. K. Cameron, C. C. Ogle, and P. D. Champion. 2004. Checklist of dicotyledons, gymnosperms, and pteridophytes naturalised or casual in New Zealand: additional records 2001–2003. New Zealand Journal of Botany **42**:797–814.

Henzel, W. J., T. M. Billeci, J. T. Stults, S. C. Wong, C. Grimley, and C. Watanabe. 1993. Identifying proteins from two-dimensional gels by molecular mass searching of peptide-fragments in protein-sequence databases. Proceedings of the National Academy of Sciences of the USA **90**:5011–5015.

Hepburn, J. S., E. Q. S. John, and F. M. Jones. 1927a. The absorption of nutrients in the pitchers of the Sarraceniaceae. Transactions of the Wagner Free Institute of Science of Philadelphia **11**:85–86.

Hepburn, J. S., and F. M. Jones. 1927. The enzymes of the pitcher liquor of the Sarraceniaceae. Transactions of the Wagner Free Institute of Science of Philadelphia **11**:49–68.

Hepburn, J. S., F. M. Jones, and E. Q. S. John. 1927b. Biochemical studies of the North American Sarraceniaceae. Transactions of the Wagner Free Institute of Science of Philadelphia **11**:1–30.

Hepburn, J. S., and E. Q. St. John. 1927. A bacteriological study of the pitcher liquor of the Sarraceniaceae. Transactions of the Wagner Free Institute of Science of Philadelphia **11**:75–83.

Hervey, J. 1747. Meditations Among the Tombs: In a Letter to a Lady. J. and J. Rivington, in St. Paul's Church-Yard; and J. Leake, at Bath, London.

Hicks, D. L., D. J. Campbell, and I. A. E. Atkinson. 2001. Options for Managing the Kaimaumau Wetland, Northland, New Zealand. Department of Conservation, Wellington, New Zealand.

Higgins, S. I., and D. M. Richardson. 1999. Predicting plant migration rates in a changing world: the role of long-distance dispersal. The American Naturalist **153**:464–475.

Hill, J. K., K. C. Hamer, J. Tangah, and M. Dawood. 2001. Ecology of tropical butterflies in rainforest gaps. Oecologia **128**:294–302.

Hillebrand, H., D. M. Bennett, and M. W. Cadotte. 2008. Consequences of dominance: a review of evenness effects on local and regional ecosystem processes. Ecology **89**:1510–1520.

Hilton, D. J. F. 1982. The biology of *Endothenia daeckeana* (Lepidoptera: Olethreutidae), an inhabitant of the ovaries of the northern pitcher plant, *Sarracenia p. purpurea* (Sarraceniaceae). The Canadian Entomologist **114**:269–274.

Hobbs, R. J., E. Higgs, and J. A. Harris. 2009. Novel ecosystems: implications for conservation and restoration. Trends in Ecology & Evolution **24**:599–605.

Hoekman, D. 2007. Top-down and bottom-up regulation in a detritus-based aquatic food web: a repeated field experiment using the pitcher plant (*Sarracenia purpurea*) inquiline community. American Midland Naturalist **157**:52–62.

Hoekman, D. 2010. Turning up the heat: temperature influences the relative importance of top-down and bottom-up effects. Ecology **91**:2819–2825.

Hoekman, D. 2011. Relative importance of top-down and bottom-up forces in food webs of *Sarracenia* pitcher communities at a northern and a southern site. Oecologia **165**:1073–1082.

Hoekman, D., R. Winston, and N. Mitchell. 2009. Top-down and bottom-up effects of a processing detritivore. Freshwater Science **28**:552–559.

Hoffman, A. A., and C. M. Sgrò. 2011. Climate change and evolutionary adaptation. Nature **470**:479–485.

Hogrefe, C., B. Lynn, K. Civerolo, J. Y. Ku, J. Rosenthal, C. Rosenzweig, et al. 2004. Simulating changes in regional air pollution over the eastern United States due to changes in global and regional climate and emissions. Journal of Geophysical Research–Atmospheres **109**:D22301.

Holdridge, L. R. 1947. Determination of world plant formations from simple climatic data. Science **105**:367–368.

Hooper, D. U., F. S. Chapin III, J. J. Ewel, A. Hector, P. Inchausti, S. Lavorel, et al. 2005. Effects of biodiversity on ecosystem functioning: a consensus of current knowledge. Ecological Monographs **75**:3–35.

Horn, H. S. 1975. Markovian properties of forest succession. In M. L. Cody and J. M. Diamond, editors. Ecology and Evolution of Communities, pp. 196–213. Harvard University Press, Cambridge, MA.

Hou, E., Y. Luo, Y. Kuang, C. Chen, X. Lu, L. Jiang, X. Luo, and D. Wen. 2020. Global meta-analysis shows pervasive phosphorus limitation of aboveground plant production in natural terrestrial ecosystems. Nature Communications **11**:637.

Houseman, G. R., G. G. Mittelbach, H. L. Reynolds, and K. L. Gross. 2008. Perturbations alter community convergence, divergence, and formation of multiple community states. Ecology **89**:2172–2180.

Huang, J. G., Y. Bergeron, B. Denneler, F. Berninger, and J. Tardif. 2007. Response of forest trees to increased atmospheric CO_2. Critical Reviews in Plant Sciences **26**:265–283.

Hubbell, S. P. 2001. The Unified Neutral Theory of Biodiversity and Biogeography. Princeton University Press, Princeton, NJ.

Hughes, J. B., J. J. Hellmann, T. H. Ricketts, and B. J. M. Bohannan. 2001. Counting the uncountable: statistical approaches to estimating microbial diversity. Applied and Environmental Microbiology **67**:4399–4406.

Human Microbiome Project Consortium. 2012. Structure, function and diversity of the healthy human microbiome. Nature **486**:207–214.

Hurlbert, S. 1971. The nonconcept of species diversity: a critique and alternative parameters. Ecology **52**:577–586.

Hurlbert, S. H. 1984. Pseudoreplication and the design of ecological field experiments. Ecological Monographs **54**:187–211.

Huston, M. A. 1984. Biological Diversity: The Coexistence of Species on Changing Landscapes. Cambridge University Press, Cambridge.

Huston, M. A. 1997. Hidden treatments in ecological experiments: re-evaluating the ecosystem function of biodiversity. Oecologia **110**:449–460.

Hutchings, J. A. 2015. Thresholds for impaired species recovery. Proceedings of the Royal Society B: Biological Sciences **282**:20150654.

Hutchinson, M. C., B. B. Mora, S. Pilosof, A. K. Barner, S. Kéfi, E. Thébault, et al. 2019. Seeing the forest for the trees: putting multilayer networks to work for community ecology. Functional Ecology 33:206–217.

IPCC. 2014. Climate Change 2014: Synthesis Report. Contribution of Working Groups I, II and III to the Fifth Assessment Report of the Intergovernmental Panel on Climate Change. IPCC, Geneva.

Isbell, F., D. Tilman, S. Polasky, S. Binder, and P. Hawthorne. 2013. Low biodiversity state persists two decades after cessation of nutrient enrichment. Ecology Letters 16:454–460.

Ives, A. R., and B. J. Cardinale. 2004. Food-web interactions govern the resistance of communities after non-random extinctions. Nature 429:174–177.

Jackson, J. B. C., L. W. Buss, and R. E. Cook, editors. 1986. Population Biology and Evolution of Clonal Organisms. Yale University Press, New Haven, CT.

Jaeglé, L., L. Steinberger, R. B. Martin, and K. Chance. 2005. Global partitioning of NO_x sources using satellite observations: relative roles of fossil fuel combustion, biomass burning and soil emissions. Faraday Discussions 130:407–423.

Janse, J. H., M. Scheffer, L. Lijklema, L. Van Liere, J. S. Sloot, and W. M. Mooij. 2010. Estimating the critical phosphorus loading of shallow lakes with the ecosystem model PCLake: sensitivity, calibration and uncertainty. Ecological Modelling 221:654–665.

Jebb, M. H. P., and M. R. Cheek. 1997. A skeletal revision of *Nepenthes* (Nepenthaceae). Blumea 42:1–106.

Jennings, D. E., A. M. Congelosi, and J. R. Rohr. 2012. Insecticides reduce survival and the expression of traits associated with carnivory of carnivorous plants. Ecotoxicology 21:569–575.

Jennings, D. E., and J. R. Rohr. 2011. A review of the conservation threats to carnivorous plants. Biological Conservation 144:1356–1363.

Joel, D. M., and S. Gepstein. 1985. Chloroplasts in the epidermis of *Sarracenia* (the American pitcher plant) and their possible role in carnivory: an immunocytochemical approach. Physiologia Plantarum 63:71–75.

Johnson, E. A., and K. Miyanishi. 2008. Testing the assumptions of chronosequences in succession. Ecology Letters 11:419–431.

Jones, F. M. 1908. Pitcher-plant insects—III. Entomological News 19:150–156.

Jones, F. M. 1916. Two insect associates of the California pitcher-plant, *Darlingtonia californica* (Dipt.). Entomological News 27:385–392.

Jones, F. M. 1920. Another pitcher-plant insect (Diptera, Sciaridae). Entomological News 31:91–95.

Jones, F. M. 1921. Pitcher plants and their moths. Natural History 21:296–316.

Jones, F. M. 1935. Pitcher plants and their insect associates. In M. V. Walcott, editor. Illustrations of North American Pitcher Plants, pp. 25–34. Smithsonian Institution, Washington, DC.

Jones, H. P., and O. J. Schmitz. 2009. Rapid recovery of damaged ecosystems. PLoS ONE 4:e5653.

Jordán, F., N. Gjata, S. Mei, and C. M. Yule. 2012. Simulating food web dynamics along a gradient: quantifying human influence. PLoS ONE 7:e40280.

Jordan, W., and G. M. Lubick. 2011. Making Nature Whole: A History of Ecological Restoration. Island Press, Washington, DC.

Judd, W. W. 1959. Studies of the Byron Bog in southwestern Ontario X. Inquilines and victims of the pitcher-plant *Sarracenia purpurea* L. The Canadian Entomologist **91**:171–180.

Juniper, B. E., R. J. Robins, and D. M. Joel. 1989. The Carnivorous Plants. Academic Press, London.

Junium, C., A. Dickson, and B. Uveges. 2018. Perturbation to the nitrogen cycle during rapid Early Eocene global warming. Nature Communications **9**:3186.

Kammenga, J. E., M. A. Herman, N. J. Ouborg, L. Johnson, and R. Breitling. 2007. Microarray challenges in ecology. Trends in Ecology & Evolution **22**:273–279.

Kanehisa, M., Y. Sato, M. Kawashima, M. Furumichi, and M. Tanabe. 2016. KEGG as a reference resource for gene and protein annotation. Nucleic Acids Research **44**:D457–D462.

Karagatzides, J. D., J. L. Butler, and A. M. Ellison. 2009. The pitcher plant *Sarracenia purpurea* can directly acquire organic nitrogen and short-circuit the inorganic nitrogen cycle. PLoS ONE **4**:e6164.

Karagatzides, J. D., and A. M. Ellison. 2009. Construction costs, payback times and the leaf economics of carnivorous plants. American Journal of Botany **96**: 1612–1619.

Karp, N. A., M. Spencer, H. Lindsay, K. O'Dell, and K. S. Lilley. 2005. Impact of replicate types on proteomic expression analysis. Journal of Proteome Research **4**:1867–1871.

Kattge, J, G. Bönisch, S. Díaz, S. Lavorel, I. C. Prentice, P. Leadley, C. Wirth, et al. 2020. TRY plant trait database—enhanced coverage and open access. Global Change Biology **26**:119–188.

Kearns, P. J., J. Angell, E. Howard, D. L.A., R. Stanley, and J. Bowen. 2016. Nutrient enrichment induces dormancy and decreases diversity of active bacteria in salt marsh sediments. Nature Communications **7**:12881.

Keddy, P. A., and B. Shipley. 1989. Competitive hierarchies in herbaceous plant communities. Oikos **54**:234–241.

Keith, D., H. R. Akçakaya, W. Thuiller, G. F. Midgley, R. G. Pearson, S. J. Phillips, et al. 2008. Predicting extinction risks under climate change: coupling stochastic population models with dynamic bioclimatic habitat models. Biology Letters **4**:560–563.

Keller, E. F. 2005. Revisiting "scale-free" networks. Bioessays **27**:1060–1068.

Kelly, A. E., and M. L. Goulden. 2008. Rapid shifts in plant distribution with recent climate change. Proceedings of the National Academy of Sciences of the USA **105**:11823–11826.

Kerkhoff, A., W. Fagan, J. Elser, and B. Enquist. 2006. Phylogenetic and growth form variation in the scaling of nitrogen and phosphorus in the seed plants. The American Naturalist **168**:E103–E122.

Kerouack, J. 1997. Some of the Dharma. Viking Penguin, New York.

Kingsolver, J. 1976. The effect of environmental uncertainty on morphological design and fluid balance in *Sarracenia purpurea* L. Oecologia **48**:364–370.

Kitching, R. L. 2000. Food Webs and Container Habitats: The Natural History and Ecology of Phytotelmata. Cambridge University Press, Cambridge.

Kneitel, J. M. 2007. Intermediate-consumer identity and resources alter a food web with omnivory. Journal of Animal Ecology **76**:651–659.

Kneitel, J. M., and T. E. Miller. 2002. Resource and top-predator regulation in the pitcher plant (*Sarracenia purpurea*) inquiline community. Ecology **83**:680–688.

Koerselman, W., and A. F. W. Meuleman. 1996. The vegetation N:P ratio: a new tool to detect the nature of nutrient limitation. Journal of Applied Ecology **33**: 1441–1450.

Koller-Peroutka, M., S. Krammer, A. Pavlik, M. Edlinger, I. Lang, and W. Adlassnig. 2019. Endocytosis and digestion in carnivorous pitcher plants of the family Sarraceniaceae. Plants **8**:367.

Komoroske, L. M., R. E. Connon, J. Lindberg, B. S. Cheng, G. Castillo, M. Hasenbein, and N. A. Fangue. 2014. Ontogeny influences sensitivity to climate change stressors in an endangered fish. Conservation Physiology **2**:cou008.

Koopman, M. M., and B. C. Carstens. 2011. The microbial phyllogeography of the carnivorous plant *Sarracenia alata*. Microbial Ecology **61**:750–758.

Koopman, M. M., D. M. Fuselier, S. Hird, and B. C. Carstens. 2010. The carnivorous pale pitcher plant harbors diverse, distinct, and time-dependent bacterial communities. Applied and Environmental Microbiology **76**:1851–1860.

Kordas, R. L., and C. D. G. Harley. 2016. Demographic responses of coexisting species to in situ warming. Marine Ecology Progress Series **546**:147–161.

Krause, A. E., K. A. Frank, D. M. Mason, R. E. Ulanowicz, and W. W. Taylor. 2003. Compartments revealed in food-web structure. Nature **426**:282–285.

Kreil, C. 1865. Klimatologie von Böhmen. Druck und Verlag von Carl Gerold's Sohn, Vienna.

Kricher, J. 2009. The Balance of Nature: Ecology's Enduring Myth. Princeton University Press, Princeton, NJ.

Krupa, S. V. 2003. Effects of atmospheric ammonia (NH_3) on terrestrial vegetation: a review. Environmental Pollution **124**:179–221.

Kuffner, T. A., and S. G. Walker. 2019. Why are *p*-values controversial? The American Statistician **73**:1–3.

Laio, F., L. Ridolfi, and P. D'Odorico. 2008. Noise-induced transitions in state-dependent dichotomous processes. Physical Review E **78**:081137.

Lallensack, R. 2018. How warp-speed evolution is transforming ecology. Nature **554**: 19–21.

Lamb, T., and E. L. Kalies. 2020. An overview of lepidopteran herbivory on North American pitcher plants (*Sarracenia*), with a novel observation of feeding on *Sarracenia flava*. The Journal of the Lepidopterists' Society **74**:193–197.

Lany, N. K., P. L. Zarnetske, E. M. Schliep, R. N. Schaeffer, C. M. Orians, D. A. Orwig, and E. L. Preisser. 2018. Asymmetric biotic interactions and abiotic niche differences revealed by a dynamic joint species distribution model. Ecology **99**:1018–1023.

Larsen, T. H., G. Brehm, H. Navarrete, P. Franco, H. Gomez, J. L. Mena, et al. 2011. Range shifts and extinctions driven by climate change in the tropical Andes: synthesis and directions. In S. K. Herzog, R. Martínez, P. M. Jørgensen, and H. Tiessen, editors. Climate Change and Biodiversity in the Tropical Andes, pp. 47–67. Inter-American Institute of Global Change Research and Scientific Committee on Problems of the Environment, Montevideo, Uruguay.

Lau, M. K., B. Baiser, N. J. Gotelli, and A. M. Ellison. 2018. Regime shifts, alternative states, and hysteresis in the *Sarracenia* microecosystem. Ecological Modelling **382**:1–8.

Lau, M. K., S. R. Borrett, B. Baiser, N. J. Gotelli, and A. M. Ellison. 2017. Ecological network metrics: opportunities for synthesis. Ecosphere **8**:e01900.

Laurel, B. J., L. A. Copeman, M. Spencer, and P. Iseri. 2017. Temperature-dependent growth as a function of size and age in juvenile arctic cod (*Boreogadus saida*). ICES Journal of Marine Science **74**:1614–1621.

Lavergne, S., N. Mouquet, W. Thuiller, and O. Ronce. 2010. Biodiversity and climate change: integrating evolutionary and ecological responses of species and communities. Annual Review of Ecology, Evolution, and Systematics **41**:321–350.

Law, R., and R. D. Morton. 1993. Alternative permanent states of ecological communities. Ecology **74**:1347–1361.

Lawton, J. H., D. E. Bignell, B. Bolton, G. F. Bloemers, P. Eggleton, P. M. Hammond, et al. 1998. Biodiversity inventories, indicator taxa and effects of habitat modification in tropical forest. Nature **391**:72–76.

Le Guin, U. K. 1969. The Left Hand of Darkness. Ace Books, New York.

Leakey, R., and R. Lewin. 1996. The Sixth Extinction: Patterns of Life and the Future of Humankind. Anchor Books, New York.

Lefkovitch, L. P. 1965. The study of population growth in organisms grouped by stages. Biometrika **35**:183–212.

Leibold, M. A., and J. M. Chase. 2017. Metacommunity Ecology. Princeton University Press, Princeton.

Leibold, M. A., and G. M. Mikkelson. 2002. Coherence, species turnover, and boundary clumping: elements of meta-community structure. Oikos **97**:237–250.

Leontief, W. 1951. The Structure of the American Economy, 1919–1939, 2nd ed. Oxford University Press, New York.

Lepetz, V., M. Massot, A. S. Chaine, and J. Clobert. 2009. Climate warming and the evolution of morphotypes in a reptile. Global Change Biology **15**:454–466.

Letcher, S. G., J. R. Lasky, R. L. Chazdon, N. Norden, S. J. Wright, et al. 2015. Environmental gradients and the evolution of successional habitat specialization: a test case with 14 Neotropical forest sites. Journal of Ecology **103**:1276–1290.

Levin, S. A. 1992. The problem of pattern and scale in ecology. Ecology **73**:1943–1967.

Levins, R., 1966. The strategy of model building in population biology. In E. Sober, editor. Conceptual Issues in Evolutionary Biology, pp. 18–27. MIT Press, Cambridge, MA.

Levins, R. 1968. Evolution in Changing Environments. Princeton University Press, Princeton, NJ.

Li, S., and C. Zhou. 2019. What are the impacts of demographic structure on CO_2 emissions? A regional analysis in China via heterogeneous panel estimates. Science of the Total Environment **650**:2021–2031.

Li, X., and X. Li. 2016. Reconstruction of stochastic temporal networks through diffusive arrival times. Nature Communications **8**:15729.

Liautaud, K., E. H. van Nes, M. Barbier, M. Scheffer, and M. Laureau. 2019. Superorganisms or loose collections of species? A unifying theory of community patterns along environmental gradients. Ecology Letters **22**:1243–1252.

Liebig, J. 1840. Die organische Chemie in ihrer Anwendung auf Agricultur und Physiologie. Friedrich Vieweg und Sohn Publ. Co., Braunschweig, Germany.

Lindeman, R. L. 1942. The trophic-dynamic aspect of ecology. Ecology **23**:399–413.

Linneaus, C. 1753. Species Plantarum. L. Salvius, Stockholm.

Litzow, M. A., J. D. Urban, and B. J. Laurel. 2008. Increased spatial variance accompanies reorganization of two continental shelf ecosystems. Ecological Applications **18**:1331–1337.

Lloyd, F. E. 1942. The Carnivorous Plants. Chronica Botanica, Waltham, MA.

Loarie, S. R., P. B. Duffy, H. Hamilton, G. P. Asner, C. B. Field, and D. D. Ackerly. 2009. The velocity of climate change. Nature 462:1052–1055.

Logan, C. A., L. E. Kost, and G. N. Somero. 2012. Latitudinal differences in *Mytilus californianus* thermal physiology. Marine Ecology Progress Series 450:93–105.

Longino, J. T., J. Coddington, and R. K. Colwell. 2002. The ant fauna of a tropical rain forest: estimating species richness three different ways. Ecology 83:689–702.

López, D. N., P. A. Camus, N. Valdivia, and S. A. Estay. 2018. Food webs over time: evaluating structural differences and variability of degree distributions in food webs. Ecosphere 9:e02539.

Lotze, H. K., and I. Milewski. 2004. Two centuries of multiple human impacts and successive changes in a North Atlantic food web. Ecological Applications 14:1428–1447.

Louizos, C., J. A. Yáñez, M. L. Forrest, and N. M. Davies. 2014. Understanding the hysteresis loop conundrum in pharmacokinetic/pharmacodynamic relationships. Journal of Pharmacy and Pharmaceutical Sciences 17:34–91.

Ludwig, D., S. R. Carpenter, and W. A. Brock. 2003. Optimal phosphorus loading for a potentially eutrophic lake. Ecological Applications 13:1135–1152.

Luis, S., C. M. Vauclair, and M. L. Lima. 2018. Raising awareness of climate change causes? Cross-national evidence for the normalization of societal risk perception of climate change. Environmental Science & Policy 80:74–81.

Lutze, J. L., J. S. Roden, C. J. Holly, J. Wolfe, J. J. G. Egerton, and M. C. Ball. 1998. Elevated atmospheric [CO_2] promotes frost damage in evergreen tree seedlings. Plant Cell and Environment 21:631–635.

Lyons, S. K., K. L. Amatangelo, A. K. Behrensmeyer, A. Bercovici, J. L. Blois, M. Davis, et al. 2016. Holocene shifts in the assembly of plant and animal communities implicate human impacts. Nature 529:80–U183.

Ma, Q., J.-G. Huang, H. Hanninen, and F. Berninger. 2019. Divergent trends in the risk of spring frost damage to trees in Europe with recent warming. Global Change Biology 25:351–360.

MacArthur, R. 1955. Fluctuations of animal populations, and a measure of community stability. Ecology 36:533–536.

MacArthur, R. H. 1971. Patterns of terrestrial bird communities. In D. S. Farner and J. R. King, editors. Avian Biology, volume 1, pp. 189–221. Academic Press, New York.

MacArthur, R. H. 1972. Geographical Ecology. Harper & Row, New York.

MacArthur, R. H., and E. O. Wilson. 1963. An equilibrium theory of insular zoogeography. Evolution 17:373–387.

Macbride, J. 1818. On the power of *Sarracenia adunca* to entrap insects. Transactions of the Linnean Society of London 12:48–52.

Maestre, F. T., R. M. Callaway, F. Valladares, and C. J. Lortie. 2009. Refining the stress-gradient hypothesis for competition and facilitation in plant communities. Journal of Ecology 97:199–205.

Maestre, F. T., J. L. Quero, N. J. Gotelli, A. Escudero, V. Ochoa, M. Delgado-Baquerizo, et al. 2012. Plant species richness and ecosystem multifunctionality in global drylands. Science 335:214–218.

Magurran, A. E., and B. J. McGill. 2011. Biological Diversity: Frontiers in Measurement and Assessment. Oxford University Press, Oxford.

Mahowald, N., T. Jickells, A. Baker, P. Artaxo, C. Benitez-Nelson, G. Bergametti, et al. 2008. Global distribution of atmospheric phosphorus sources, concentrations and deposition rates, and anthropogenic impacts. Global Biogeochemical Cycles 22:GB4026.

Malis, F., M. Kopecky, P. Petrik, J. Vladovic, J. Merganic, and T. Vida. 2016. Life stage, not climate change, explains observed tree range shifts. Global Change Biology 22:1904–1914.

Malmberg, R. L., W. L. Rogers, and M. S. Alabady. 2018. A carnivorous plant genetic map: pitcher/insect-capture QTL on a genetic linkage map of *Sarracenia*. Life Science Alliance 1:e201800146.

Malthus, T. R. 1798. An Essay on the Principle of Population As It Affects the Future Improvement of Society. J. Johnson, in St. Paul's Church-yard, London.

Mandossian, A. 1966. Variations in the leaf of *Sarracenia purpurea* (pitcher plant). The Michigan Botanist 5:26–35.

Mandossian, A. J. 1965. *Some Aspects of the Ecological Life History of* Sarracenia purpurea. PhD dissertation, Michigan State University.

Manning, K., and A. Timpson. 2014. The demographic response to Holocene climate change in the Sahara. Quaternary Science Reviews 101:28–35.

Mantua, N. J., and S. R. Hare. 2002. The Pacific decadal oscillation. Journal of Oceanography 58:35–44.

Martinez, N. D. 1992. Constant connectance in community food webs. The American Naturalist 139:1208–1218.

Marx, R. 1914. Maîtres d'hier et d'aujourd'hui. Calmann-Lévy, Éditeurs, Paris.

Mattson, W. J. 1980. Herbivory in relation to plant nitrogen content. Annual Reviews of Ecology and Systematics 11:119–161.

May, R. M. 1972. Will a large complex system be stable? Nature 238:413–414.

May, R. M. 1977. Thresholds and breakpoints in ecosystems with a multiplicity of stable states. Nature 269:471–477.

May, R. M. 2006. Network structure and the biology of populations. Trends in Ecology & Evolution 21:394–399.

May, R. M., and A. R. McLean, editors. 2007. Theoretical Ecology: Principles and Applications. 3rd ed. Oxford University Press, Oxford.

McAuliffe, J. R. 1988. Markovian dynamics of simple and complex desert plant communities. The American Naturalist 131:459–490.

McCann, K., A. Hastings, and G. R. Huxel. 1998. Weak trophic interactions and the balance of nature. Nature 395:794–798.

McCann, K. S. 2012. Food Webs. Princeton University Press, Princeton.

McCoy, E. D., and K. L. Heck. 1987. Some observations on the use of taxonomic similarity in large-scale biogeography. Journal of Biogeography 14:79–87.

McDaniel, S. 1971. The genus *Sarracenia* (Sarraceniaceae). Bulletin of the Tall Timbers Research Station 9:1–36.

McKenney, D. W., J. H. Pedlar, K. Lawrence, P. Papadopol, K. Campbell, and M. F. Hutchinson. 2014. Change and evolution in the plant hardiness zones of Canada. BioScience 64:341–350.

McPherson, S., and D. Schnell. 2011. Sarraceniaceae of North America. Redfern Natural History Productions, Poole.

Meier, M., J. Fuhrer, and A. Holzkämper. 2018. Changing risk of spring frost damage in grapevines due to climate change? A case study in the Swiss Rhone Valley. International Journal of Biometeorology 62:991–1002.

Mellichamp, J. H., 1875. Notes on *Sarracenia variolaris*. In F. W. Putnam, editor. Proceedings of the American Association for the Advancement of Science, Twenty-Third Meeting, held at Hartford Connecticut; Part II. B. Natural History, pp. 113–133. The Salem Press, Salem.

Mellichamp, T. L., and F. W. Case, 2009. *Sarracenia*. In F. of North America Editorial Committee, editor. Flora of North America, volume 8, pp. 350–363. Oxford University Press, New York.

Méndez-García, C., A. I. Peláez, V. Mesa, J. Sánchez, O. Golyshina, and M. Ferrer. 2015. Microbial diversity and metabolic networks in acid mine drainage habitats. Frontiers in Microbiology **6**:475.

Menge, B. A. 1995. Indirect effects in marine rocky intertidal interaction webs: patterns and importance. Ecology Letters **65**:21–74.

Menge, B. A. 2000. Top-down and bottom-up community regulation in marine rocky intertidal habitats. Journal of Experimental Marine Biology and Ecology **250**: 257–289.

Merz, C., J. M. Catchen, V. Hanson-Smith, K. J. Emerson, W. E. Bradshaw, and C. M. Holzapfel. 2013. Replicate phylogenies and post-glacial range expansion of the pitcher-plant mosquito, *Wyeomyia smithii*, in North America. PLoS ONE **8**:e72262.

Middelburg, J. J., and L. A. Levin. 2009. Coastal hypoxia and sediment biogeochemistry. Biogeosciences **6**:1273–1293.

Miller, T. E., W. E. Bradshaw, and C. M. Holzapfel, 2018. Pitcher-plant communities as model systems for addressing fundamental questions in ecology and evolution. In A. M. Ellison and L. Adamec, editors. Carnivorous Plants: Physiology, Ecology, and Evolution, pp. 333–348. Oxford University Press, Oxford.

Miller, T. E., and J. M. Kneitel, 2005. Inquiline communities in pitcher plants as a prototypical metacommunity. In M. Holyoak, M. A. Leibold, and R. D. Holt, editors. Metacommunities: Spatial Dynamics and Ecological Communities, pp. 122–145. University of Chicago Press, Chicago.

Miller, T. E., and C. P. terHorst. 2012. Testing successional hypotheses of stability, heterogeneity, and diversity in pitcher-plant inquiline communities. Oecologia **170**:243–251.

Milne, M. A. 2012. The purple pitcher plant as a spider oviposition site. Southeastern Naturalist **11**:567–574.

Milne, M. A., and D. A. Waller. 2018. Carnivorous pitcher plants eat a diet of certain spiders, regardless of what's on the menu. Ecosphere **9**:e02504.

Mittelbach, G. G., E. A. Garcia, and Y. Taniguchi. 2006. Fish reintroductions reveal smooth transitions between lake community states. Ecology **87**:312–318.

Mladenov, N., M. Williams, S. Schmidt, and K. Cawley. 2012. Atmospheric deposition as a source of carbon and nutrients to an alpine catchment of the Colorado Rocky Mountains. Biogeosciences **9**:3337–3355.

Mlynarek, J. J., and T. A. Wheeler. 2018. Chloropid flies (Diptera, Chloropidae) associated with pitcher plants in North America. PeerJ **6**:e4491.

Moldowan, P. D., M. A. Smith, T. Baldwin, T. Bartley, N. Rollinson, and H. Wynen. 2019. Nature's pitfall trap: salamanders as rich prey for carnivorous plants in a nutrient-poor northern bog ecosystem. Ecology **100**:e02770.

Moore, T. R., J. L. Bubier, S. E. Frolking, P. M. Lafleur, and N. T. Roulet. 2002. Plant biomass and production and CO_2 exchange in an ombrotrophic bog. Journal of Ecology **90**:25–36.

Mora, B. B., D. Gravel, L. J. Gilarranz, T. Poisot, and D. B. Stouffer. 2018. Identifying a common backbone of interactions underlying food webs from different ecosystems. Nature Communications **9**:2603.

Morlon, H., E. P. White, R. S. Etienne, J. L. Green, A. Ostling, D. Alonso, et al. 2009. Taking species abundance distributions beyond individuals. Ecology Letters **12**:488–501.

Morton, R. D., R. Law, S. L. Pimm, and J. Drake. 1996. On models for assembling ecological communities. Oikos **75**:493–499.

Mouquet, N., T. Daufresne, S. M. Gray, and T. E. Miller. 2008. Modelling the relationship between a pitcher plant (*Sarracenia purpurea*) and its phytotelma community: mutualism or parasitism? Functional Ecology **22**:728–737.

Mueller, N. D., J. S. Gerber, M. Johnston, D. K. Ray, N. Ramankutty, and J. A. Foley. 2012. Closing yield gaps through nutrient and water management. Nature **490**: 254–257.

Müller, F. 1997. State-of-the-art in ecosystem theory. Ecological Modelling **100**:135–161.

Murren, C., J. Auld, H. Callahan, C. Ghalambor, C. Handelsman, M. Heskel, et al. 2015. Constraints on the evolution of phenotypic plasticity: limits and costs of phenotype and plasticity. Heredity **115**:293–301.

Naczi, R. F. C. 2018. Systematics and evolution of Sarraceniaceae. In A. M. Ellison and L. Adamec, editors. Carnivorous Plants: Physiology, Ecology, and Evolution, pp. 105–119. Oxford University Press, Oxford.

Naczi, R. F. C., E. M. Soper, F. W. Case, and R. B. Case. 1999. *Sarracenia rosea* (Sarraceniaceae), a new species of pitcher plant from the southeastern United States. Sida **18**:1183–206.

Naeem, S. 1988. Resource heterogeneity fosters coexistence of a mite and a midge in pitcher plants. Ecological Monographs **58**:215–227.

NCBI Resource Coordinators. 2017. Database resources of the National Center for Biotechnology. Nucleic Acids Research **45**:D12–D17.

Ne'eman, G., R. Ne'eman, and A. M. Ellison. 2006. Limits to reproductive success of *Sarracenia purpurea* (Sarraceniaceae). American Journal of Botany **93**: 1660–1666.

Nesatyy, V. J., and M. J.-F. Suter. 2007. Proteomics for the analysis of environmental stress responses in organisms. Environmental Science & Technology **41**:6891–6900.

Newell, S. J., and A. J. Nastase. 1998. Efficiency of insect capture by *Sarracenia purpurea* (Sarraceniaceae), the northern pitcher plant. American Journal of Botany **85**:88–91.

Neyland, R., and M. Merchant. 2006. Systematic relationships of Sarraceniaceae inferred from nuclear ribosomal DNA sequences. Madroño **53**:223–232.

Ngai, J. T., and D. S. Srivastava. 2006. Predators accelerate nutrient cycling in a bromeliad ecosystem. Science **314**:963–963.

Nielsen, D. W. 1990. Arthropod communities associated with *Darlingtonia californica*. Annals of the Entomological Society of America **83**:189–200.

NOAA, 2018. Global greenhouse gase reference network: trends in atmospheric carbon dioxide. https://www.esrl.noaa.gov/gmd/ccgg/trends/full.html.

Nogués-Bravo, D. 2009. Predicting the past distribution of species climatic niches. Global Ecology and Biogeography **18**:521–531.

Northrop, A. C., V. Avalone, A. M. Ellison, B. A. Ballif, and N. J. Gotelli. 2021. Clockwise and counterclockwise hysteresis characterize state changes in the same aquatic ecosystem. Ecology Letters **24**:94–101.

Northrop, A. C., R. Brooks, A. M. Ellison, N. J. Gotelli, and B. A. Ballif. 2017. Metaproteomics reveals taxonomic and functional changes in an enriched aquatic ecosystem? Ecosphere **8**:e01954.

O'Connor, F. 1969. The nature and aim of fiction. In S. Fitzgerald and R. Fitzgerald, editors. Mystery and Manners: Occasional Prose, pp. 63–86. Farrar, Straus & Giroux, New York.

Odion, D. C., M. A. Moritz, and D. A. DellaSala. 2010. Alternative community states maintained by fire in the Klamath Mountains, USA. Journal of Ecology **98**: 96–105.

OED Online, 2018a. "community, n." http://www.oed.com/view/Entry/37337.

OED Online, 2018b. "scale, n1–n7." http://www.oed.com/view/Entry/171737.

Oelschlägel, B., S. Gorb, S. Wanke, and C. Neinhuis. 2009. Structure and biomechanics of trapping flower trichomes and their role in the pollination biology of *Aristolochia* plants (Aristolochiaceae). New Phytologist **184**:988–1002.

Oelschlägel, B., M. Nuss, M. von Tschirnhaus, C. Pätzold, C. Neinhuis, S. Dötterl, and S. Wanke. 2015. The betrayed thief: the extraordinary strategy of *Aristolochia rotunda* to deceive its pollinators. New Phytologist **206**:342–351.

O'Gorman, E. J., and M. C. Emmerson. 2009. Perturbations to trophic interactions and the stability of complex food webs. Proceedings of the National Academy of Sciences of the USA **106**:13393–13398.

Ohlmann, M., V. Miele, S. Dray, L. Chalmandrier, L. O'Connor, and W. Thuiller. 2019. Diversity indices for ecological networks: a unifying framework using Hill numbers. Ecology Letters **22**:737–747.

Olde Venterink, H., N. M. Pieterse, J. D. M. Belgers, M. J. Wassen, and P. C. De Ruiter. 2002. N, P, and K budgets along nutrient availability and productivity gradients in wetlands. Ecological Applications **12**:1010–1026.

Olde Venterink, H., M. J. Wassen, A. W. M. Verkroost, and P. C. De Ruiter. 2003. Species richness–productivity patterns differ between N-, P-, and K-limited wetlands. Ecology **84**:2191–2199.

Olesen, J. M., C. Stefanescu, and A. Traveset. 2011. Strong, long-term temporal dynamics of an ecological network. PLoS ONE **6**:e26455.

Olson, D. M., E. Dinerstein, E. D. Wikramanayake, N. D. Burgess, G. V. N. Powell, E. C. Underwood, et al. 2001. Terrestrial ecoregions of the world: a new map of life on Earth. BioScience **51**:933–938.

Olszewski, T. D. 2004. A unified mathematical framework for the measurement of richness and evenness within and among multiple communities. Oikos **104**: 377–387.

Osnas, J., M. Katabuchi, K. Kitajim, S. Wright, P. Reich, S. van Bael, et al. 2018. Divergent drivers of leaf trait variation within species, among species, and among functional groups. Proceedings of the National Academy of Sciences of the USA **115**:5480–5485.

Östman, O., N. W. Griffin, J. L. Strasburg, J. A. Brisson, A. R. Templeton, T. M. Knight, and J. M. Chase. 2007. Habitat area affects arthropod communities directly and indirectly through top predators. Ecography **30**:359–366.

Pacala, S. W., C. D. Canham, J. Saponara, J. A. Silander, R. K. Kobe, and E. Ribbens. 1996. Forest models defined by field measurements: estimation, error analysis and dynamics. Ecological Monographs **66**:1–43.

Pace, M. L., S. R. Carpenter, and J. J. Cole. 2015. With and without warning: managing ecosystems in a changing world. Frontiers in Ecology and the Environment **13**:460–467.

Paine, R. T. 1966. Food web complexity and species diversity. The American Naturalist **100**:65–75.

Paine, R. T. 1988. Food webs: road maps of interactions or grist for theoretical development. Ecology **69**:1648–1654.

Paine, R. T. 1992. Food-web analysis through field measurement of per-capita interaction strength. Nature **355**:73–75.

Paine, R. T. 2002. Trophic control of production in a rocky intertidal community. Science **296**:736–739.

Paise, T. K., T. E. Miller, and O. U. Mason. 2014. Effects of a ciliate protozoa predator on microbial communities in pitcher plant *Sarracenia purpurea* leaves. PLoS ONE **9**:e113384.

Parain, E. C., S. M. Gray, and L.-F. Bersier. 2019. The effects of temperature and dispersal on species diversity in natural microbial metacommunities. Scientific Reports **9**:18286.

Pardo, L., M. Robin-Abbott, and C. Driscoll. 2010. Assessment of Nitrogen Deposition Effects and Empirical Critical Loads of Nitrogen for Ecoregions of the United States. General Technical Report NRS-80. United States Department of Agriculture Northern Research Station, Newtown Square, PA.

Parisod, C., C. Tripi, and N. Galland. 2005. Genetic variability and founder effect in the pitcher plant *Sarracenia purpurea* (Sarraceniaceae) in populations introduced into Switzerland: from inbreeding to invasion. Annals of Botany **95**:277–286.

Park, D. S., I. Breckheimer, A. C. Williams, E. Law, A. M. Ellison, and C. C. Davis. 2018. Herbarium specimens reveal substantial and unexpected variation in phenological sensitivity across the eastern United States. Philosophical Transactions of the Royal Society B: Biological Sciences **274**:20170394.

Pascal, B. 1897. Pensées et Opuscules. Hachette, Paris.

Pascual, M., and F. Guichard. 2005. Criticality and disturbance in spatial ecological systems. Trends in Ecology & Evolution **20**:88–95.

Patten, B. C. 1959. An introduction to the cybernetics of the ecosystem: the trophic-dynamic aspect. Ecology **40**:221–231.

Patten, B. C., and E. P. Odum. 1981. The cybernetic nature of ecosystems. The American Naturalist **118**:886–895.

Patterson, B. D., and W. Atmar. 1986. Nested subsets and the structure of insular mammalian faunas and archipelagoes. Biological Journal of the Linnean Society **28**:65–82.

Pauly, D., V. Christensen, J. Dalsgaard, R. Froese, and F. Torres. 1998. Fishing down marine food webs. Science **279**:860–863.

Pauly, D., R. Froese, and M. L. Palomares. 2000. Fishing down aquatic food webs. American Scientist **88**:46–51.

Pavlovič, A., E. Masarovičová, and J. Hudák. 2007. Carnivorous syndrome in Asian pitcher plants of the genus *Nepenthes*. Annals of Botany **100**:527–536.

Paytan, A., and K. McLaughlin. 2007. The oceanic phosphorus cycle. Chemical Reviews **107**:563–576.

Pena, P., and M. de L'Obel. 1576. Nova Stirpium Adversaria, Perfacilis Vestigatio: Luculenta que Accessio ad Priscorum, Praesertim Dioscoridis, & Recentiorum, Materiam Medicam. C. Plantinum, Antwerp.

Persson, J., P. Fink, A. Goto, J. M. Hood, J. Jonas, and S. Kato. 2010. To be or not to be what you eat: regulation of stoichiometric homeostasis among autotrophs and heterotrophs. Oikos **119**:741–751.

Petermann, J. S., V. F. Farjalla, M. Jocque, P. Kratina, A. A. M. MacDonald, N. A. C. Marino, et al. 2015a. Dominant predators mediate the impact of habitat size on trophic structure in bromeliad invertebrate communities. Ecology **96**:428–439.

Petermann, J. S., P. Kratina, N. A. C. Marino, A. A. M. MacDonald, and D. S. Srivastava. 2015b. Resources alter the structure and increase stochasticity in bromeliad microfauna communities. PLoS ONE **10**:e0118952.

Peterson, A. T., J. Soberón, R. G. Pearson, R. P. Anderson, E. Martínez-Meyer, M. Naka-mura, and M. B. Araújo. 2011. Ecological Niches and Geographic Distributions. Princeton University Press, Princeton, NJ.

Peterson, C. N., S. Day, B. E. Wolfe, A. M. Ellison, R. Kolter, and A. Pringle. 2008. A keystone predator controls bacterial diversity in the pitcher plant (*Sarracenia purpurea*) microecosystem. Environmental Microbiology **10**:2257–2266.

Petit, J. R., J. Jouzel, D. Raynaud, N. I. Barkov, J. M. Barnola, I. Basile, et al. 1999. Climate and atmospheric history of the past 420,000 years from the Vostok ice core, Antarctica. Nature **399**:429–436.

Petraitis, P. 2013. Multiple Stable States in Natural Ecosystems. Oxford University Press, New York.

Petraitis, P. S., and R. E. Latham. 1999. The importance of scale in testing the origins of alternative community states. Ecology **80**:429–442.

Petraitis, P. S., E. T. Methratta, E. C. Rhile, N. A. Vidergas, and S. R. Dudgeon. 2009. Experimental confirmation of multiple community states in a marine ecosystem. Oecologia **161**:139–148.

Phillips, S. J., R. P. Anderson, and R. E. Schapire. 2006. Maximum entropy modeling of species geographic distributions. Ecological Modelling **190**:231–259.

Pianka, E. R. 1970. On *r*- and *K*-selection. The American Naturalist **104**:592–597.

Pianka, E. R. 1973. The structure of lizard communities. Annual Review of Ecology and Systematics **4**:53–74.

Pickett, S. T. A., and P. S. White, editors. 1985. The Ecology of Natural Disturbance and Patch Dynamics. Academic Press, Orlando.

Pielou, D. P., and E. C. Pielou. 1968. Associations among species of infrequent occurrence: the insect and spider fauna of *Polyporus betulinus* (Bulliard) Fries. Journal of Theoretical Biology **21**:202–216.

Pielou, E. C. 1972. 2^k contingency tables in ecology. Journal of Theoretical Biology **34**:337–352.

Pillai, P., and T. C. Gouhier. 2019. Not even wrong: the spurious measurement of biodiversity's effects on ecosystem functioning. Ecology **100**:e02645.

Pilosof, S., P. M.A., P. M., and S. Kéfi. 2017. The multilayer nature of ecological networks. Nature Ecology and Evolution **1**:0101.

Pimm, S. L. 1982. Food Webs. Chapman and Hall, London.

Piovia-Scott, J., L. H. Yang, and A. N. Wright. 2017. Temporal variation in trophic cascades. Annual Review of Ecology, Evolution, and Systematics **48**:281–300.

Płachno, B. J., and L. E. Muravnik, 2018. Functional anatomy of carnivorous traps. In A. M. Ellison and L. Adamec, editors. Carnivorous Plants: Physiology, Ecology, and Evolution, pp. 167–179. Oxford University Press, Oxford.

Płachno, B. J., P. Świątek, and A. Wistuba. 2007. The giant extra-floral nectaries of carnivorous *Heliamphora folliculata*: architecture and ultrastructure. Acta Biologica Cracoviensia Series Botanica **49**:91–104.

Platt, W. J., and J. H. Connell. 2003. Natural disturbances and directional replacement of species. Ecological Monographs **73**:507–522.

Poff, N. L., J. D. Allan, M. B. Bain, J. R. Karr, K. L. Prestegaard, B. D. Richter, et al. 1997. The natural flow regime. BioScience **47**:769–785.

Poff, N. L., J. D. Olden, N. K. M. Vieira, D. S. Finn, M. P. Simmons, and B. C. Kondratieff. 2006. Functional trait niches of North American lotic insects: traits-based ecological applications in light of phylogenetic relationships. Journal of the North American Benthological Society **25**:730–755.

Poisot, T., E. Canard, D. Mouillot, N. Mouquet, and D. Gravel. 2012. The dissimilarity of species interaction networks. Ecology Letters **15**:1353–1361.

Poisot, T., D. B. Stouffer, and D. Gravel. 2015. Beyond species: why ecological interaction networks vary through space and time. Oikos **124**:234–251.

Polis, G. A. 1991. Complex desert food webs: an empirical critique of food web theory. The American Naturalist **138**:123–155.

Pollock, L. J., R. Tingley, W. K. Morris, N. Golding, R. B. O'Hara, K. M. Parris, et al. 2014. Understanding co-occurrence by modelling species simultaneously with a joint species distribution model (JSDM). Methods in Ecology and Evolution **5**:397–406.

Poorter, H., S. Pepin, T. Rijkers, Y. de Jong, J. R. Evans, and C. Körner. 2006. Construction costs, chemical composition and payback time of high- and low-irradiance leaves. Journal of Experimental Botany **57**:355–371.

Poorter, H., and R. Villar, 1997. The fate of acquired carbon in plants: chemical composition and construction costs. In F. A. Bazzaz and J. Grace, editors. Plant Resource Allocation, pp. 39–72. Academic Press, San Diego.

Poulson, T. L., and W. J. Platt. 1996. Replacement patterns of beech and sugar maple in Warren Woods, Michigan. Ecology **77**:1243–1253.

Pringle, R. M., T. R. Kartzinel, T. M. Palmer, T. J. Thurman, K. Fox-Dobbs, C. C. Y. Xu, et al. 2019. Predator-induced collapse of niche structure and species coexistence. Nature **570**:58–64.

Prism Climate Group, 2004. PRISM gridded climate data. http://prism.oregonstate.edu.

Rabinowitz, D., 1981. Seven forms of rarity. In H. Synge, editor. The Biological Aspects of Rare Plant Conservation, pp. 205–216. John Wiley & Sons, Chichester, UK.

Ram, R. J., N. C. Verberkmoes, M. P. Thelen, G. W. Tyson, B. J. Baker, R. C. I. Blake, et al. 2005. Community proteomics of a natural microbial biofilm. Science **308**:1915–1920.

Ratajczak, Z., S. R. Carpenter, A. R. Ives, C. J. Kucharik, T. Ramiadantsoa, M. A. Stegner, et al. 2018. Abrupt change in ecological systems: inference and diagnosis. Trends in Ecology & Evolution **33**:513–526.

Rayburn, J. A., D. Franzi, and P. L. K. Knuepfer. 2007. Evidence from the Lake Champlain Valley for a later onset of the Champlain Sea and implications for late glacial meltwater routing to the North Atlantic. Palaeogeography, Palaeoclimatology, Palaeoecology **246**:62–74.

Razon, L. 2018. Reactive nitrogen: a perspective on its global impact and prospects for its sustainable production. Sustainable Production and Consumption **15**:35–48.

Record, S., M. C. Fitzpatrick, A. O. Finley, S. Veloz, and A. M. Ellison. 2013. Should species distribution models account for spatial autocorrelation? A test of model projections across eight millenia of climate change. Global Ecology and Biogeography **22**:760–771.

Reich, P. B. 2014. The world-wide 'fast-slow' plant economics spectrum: a traits manifesto. Journal of Ecology **102**:275–301.

Reich, P. B., and J. Oleksyn. 2004. Global patterns of plant leaf N and P in relation to temperature and latitude. Proceedings of the National Academy of Sciences of the USA **101**:11001–11006.

Reid, C. 1899. The Origin of the British Flora. Dulau, London.

Reusch, T. B. H., and T. E. Wood. 2007. Molecular ecology of global change. Molecular Ecology **16**:3973–3992.

Reveal, J. L. 1993. The correct name of the northern expression of *Sarracenia purpurea* L. (Sarraceniaceae). Phytologia **74**:180–184.

Reznick, D. N., J. Losos, and J. Travis. 2019. From low to high gear: there has been a paradigm shift in our understanding of evolution. Ecology Letters **22**: 233–244.

Richardson, B. A., M. J. Richardson, F. N. Scatena, and W. H. McDowell. 2000. Effects of nutrient availability and other elevational changes on bromeliad populations and their invertebrate communities in a humid tropical forest in Puerto Rico. Journal of Tropical Ecology **16**:167–188.

Ricklefs, R. E. 2003. A comment on Hubbell's zero-sum ecological drift model. Oikos **100**:185–192.

Ricklefs, R. E. 2008. Disintegration of the ecological community. The American Naturalist **172**:741–750.

Ridd, P. V., E. T. da Silva, and T. Stieglitz. 2013. Have coral calcification rates slowed in the last twenty years?). Marine Geology **346**:392–399.

Roberts, A. 1974. The stability of a feasible random ecosystem. Nature **251**:607–608.

Rockström, J., W. Steffen, K. Noone, A. Persson, F. S. I. Chapin, E. Lambin, et al. 2009. Planetary boundaries: exploring the safe operating space for humanity. Ecology and Society **14**:32.

Rodionov, S. N., 2005a. A brief overview of the regime shift detection methods. In V. Velikova and N. Chipev, editors. Large-scale Disturbances (Regime Shifts) and Recovery in Aquatic Ecosystems: Challenges for Management Toward Sustainability, pp. 17–24 . UNESCO-ROSTE/BAS Workshop on Regime Shifts, Varna, Bulgaria.

Rodionov, S. N. 2005b. A sequential algorithm for testing climate regime shifts. Geophysical Research Letters **31**:L09204.

Rogers, H. H., G. B. Runion, and S. V. Krupa. 1994. Plant responses to atmospheric CO_2 enrichment with emphasis on roots and the rhizosphere. Environmental Pollution **83**:155–189.

Romero, G. Q., G. C. O. Piccoli, P. M. de Omena, and T. Gonçalves-Souza. 2016. Food web structure shaped by habitat size and climate across a latitudinal gradient. Ecology **97**:2705–2715.

Rooney, N., and K. S. McCann. 2012. Integrating food web diversity, structure and stability. Trends in Ecology & Evolution **27**:40–46.

Root, R. B. 1967. The niche exploitation pattern of the blue-gray gnatcatcher. Ecological Monographs **37**:317–350.

Rose, K. C., R. A. Graves, W. D. Hansen, B. J. Harvey, J. Ziu, S. A. Wood, et al. 2017. Historical foundations and future directions in macrosystems ecology. Ecology Letters **20**:147–157.

Royle, J. A., R. B. Chandler, C. Yackulic, and J. D. Nichols. 2012. Likelihood analysis of species occurrence probability from presence-only data for modelling species distributions. Methods in Ecology and Evolution **3**:545–554.

Royle, J. A., and R. M. Dorazio. 2008. Hierarchical Modeling and Inference in Ecology: The Analysis of Data from Populations, Metapopulations, and Communities. Academic Press, London.

Rutledge, R. W., B. L. Basorre, and R. J. Mulholland. 1976. Ecological stability: an information theory viewpoint. Journal of Theoretical Biology **57**:355–371.

Rymal, D. E., and G. W. Folkerts. 1982. Insects associated with pitcher plant (*Sarracenia*: Sarraceniaceae), and their relationship to pitcher plant conservation: a review. Journal of the Alabama Academy of Science **53**:131–151.

Salguero-Gómez, R., O. R. Jones, C. R. Archer, Y. M. Buckley, J. Che-Castaldo, H. Caswell, et al. 2015. The COMPADRE plant matrix database: an open online repository for plant demography. Journal of Ecology **103**:202–218.

Sanderson, J. G. 2000. Testing ecological patterns. American Scientist **88**:332–339.

Sardens, J., and J. Peñuelas. 2012. The role of plants in the effects of global change on nutrient availability and stoichiometry in the plant-soil system. Plant Physiology **160**:1741–1761.

Sarwar, M., P. Rodriguez, and C.-Z. Li. 2019. Sweat-based in vitro diagnostics (IVD): from sample collection to point-of-care testing (POCT). Journal of Analysis and Testing **3**:80–88.

Satler, J. D., A. J. Zellmer, and B. C. Carstens. 2016. Biogeographic barriers drive co-diversification within associated eukaryotes of the *Sarracenia alata* pitcher plant system. PeerJ **4**:e1576.

Schaberg, P., D. DeHayes, G. Hawley, G. Strimbeck, J. Cumming, P. Murakami, and C. Borer. 2000. Acid mist and soil Ca and Al alter the mineral nutrition and physiology of red spruce. Tree Physiology **20**:73–85.

Scheffer, M. 2009. Critical Transitions in Nature and Society. Princeton University Press, Princeton, NJ.

Scheffer, M., J. Bascompte, W. A. Brock, V. Brovkin, S. R. Carpenter, V. Dakos, et al. 2009. Early-warning signals for critical transitions. Nature **461**:53–59.

Scheffer, M., and S. R. Carpenter. 2003. Catastrophic regime shifts in ecosystems: linking theory to observation. Trends in Ecology & Evolution **18**:648–656.

Schenk, H. J., and R. B. Jackson. 2002. Rooting depths, lateral root spreads and below-ground/above-ground allometries of plants in water-limited ecosystems. Journal of Ecology **90**:480–494.

Schiaffino, M. R., N. Diovisalvi, D. Marfetán Molina, C. Li Puma, L. Lagomarsino, M. V. Quiroga, and G. L. Pérez. 2019. Microbial food-web components in two hypertrophic human-impacted Pampean shallow lakes: interactive effects of environmental, hydrological, and temporal drivers. Hydrobiologia **830**:255–276.

Schimel, D. S. 1995. Terrestrial ecosystems and the carbon cycle. Global Change Biology **1**:77–91.

Schindler, D. W. 2012. The dilemma of controlling cultural eutrophication of lakes. Proceedings of the Royal Society B: Biological Sciences **279**:4322–4333.

Schmitt, J., and R. D. Wulff. 1993. Light spectral quality, phytochrome and plant competition. Trends in Ecology & Evolution **8**:47–51.

Schmitz, O. J. 1994. Resource edibility and trophic exploitation in an old-field food web. Proceedings of the National Academy of Sciences of the USA **91**:5364–5367.

Schmitz, O. J. 2003. Top predator control of plant diversity and productivity in an old field ecosystem. Ecology Letters **6**:156–163.

Schmitz, O. J. 2010. Resolving Ecosystem Complexity. Princeton University Press, Princeton, NJ.

Schneider, T., and K. Riedel. 2010. Environmental proteomics: analysis of structure and function of microbial communities. Proteomics **10**:785–798.

Schnell, D. E. 1978. *Sarracenia* L. petal extract chromatography. Castanea **43**:107–115.

Schnell, D. E. 1979. A critical review of published variants of *Sarracenia purpurea* L. Castanea **44**:47–59.

Schnell, D. E. 1993. *Sarracenia purpurea* L. ssp. *venosa* (Raf.) Wherry var. *burkii* Schnell (Sarraceniaceae): a new variety of the Gulf coastal plain. Rhodora **95**:6–10.

Schnell, D. E. 2002. Carnivorous Plants of the United States and Canada. Timber Press, Portland, OR.

Schnell, D. E., and R. O. Determann. 1997. *Sarracenia purpurea* L. ssp. *venosa* (Raf.) Wherry var. *montana* Schnell & Determann (Sarraceniaceae): a new variety. Castanea **62**:60–62.

Schoener, T. W. 1974. Resource partitioning in ecological communities. Science **185**:27–39.

Schröder, A., L. Persson, and A. M. de Roos. 2005. Direct experimental evidence for alternative stable states: a review. Oikos **110**:3–19.

Schröter, M., E. H. van der Zanden, A. P. E. van Oudenhoven, R. P. Remme, H. M. Serna-Chavez, R. S. de Groot, and P. Opdam. 2014. Ecosystem services as a contested concept: a synthesis of critique and counter-arguments. Conservation Letters **7**: 514–523.

Schuur, E. A. G., and M. C. Mack. 2018. Ecological response to permafrost thaw and consequences for local and global ecosystem services. Annual Review of Ecology, Evolution, and Systematics **49**:279–301.

Schwaegerle, K. E. 1983. Population growth of the pitcher plant, *Sarracenia purpurea* L., at Carnberry Bog, Licking County, Ohio. Ohio Journal of Science **83**:19–22.

Schwaegerle, K. E., and B. A. Schaal. 1979. Genetic variability and founder effects in the pitcher plant *Sarracenia purpurea*. Evolution **33**:1210–1218.

Sequeira, A. M. M., P. J. Bouchet, K. L. Yates, K. Mengersen, and M. J. Caley. 2018. Transferring biodiversity models for conservation: opportunities and challenges. Methods in Ecology and Evolution **9**:1250–1264.

Sexton, W. K., M. Fidero, J. C. Spain, L. Jiang, K. Bucalo, J. M. Cruse-Sanders, and G. S. Pullman. 2020. Characterization of endopyhtic bacterial communities within greenhouse and field-grown rhizomes of three rare pitcher plant species (*Sarracenia oreophila, S. leucophylla,* and *S. purpurea ssp. venosa*) with an emphasis on nitrogen-fixing bacteria. Plant and Soil **447**:257–279.

Sfenthourakis, S., E. Tzanatos, and S. Giokas. 2006. Species co-occurrence: the case of congeneric species and a causal approach to patterns of species association. Global Ecology and Biogeography **15**:39–49.

Shade, A., R. Dunn, S. Blowes, P. Keil, B. Bohannan, M. Herrmann, et al. 2018. Macroecology to unite all live, large and small. Trends in Ecology & Evolution **33**:731–744.

Shen, Y., J. M. Jacobs, D. G. Camp, R. Fang, R. J. Moore, R. D. Smith, et al. 2004. Ultra-high-efficiency strong cation exchange LC/RPLC/MS/MS for high dynamic range characterization of the human plasma proteome. Analytical Chemistry **76**:1134–1144.

Sheridan, P. M., and D. N. Karowe. 2000. Inbreeding, outbreeding, and heterosis in the yellow pitcher plant, *Sarracenia flava* (Sarraceniaceae) in Virginia. American Journal of Botany **87**:1628–33.

Sheridan, P. M., and R. R. Mills. 1998. Genetics of anthocyanin deficiency in *Sarracenia* L. Hortscience **33**:1042–45.

Shreve, F. 1906. The development and anatomy of *Sarracenia purpurea*. The Botanical Gazette **42**:107–126.

Shurin, J. B., D. S. Gruner, and H. Hillebrand. 2006. All wet or dried up? Real differences between aquatic and terrestrial food webs. Proceedings of the Royal Society B: Biological Sciences **273**:1–9.

Sibly, R. M., and J. Hone. 2002. Population growth rate and its determinants: an overview. Philosophical Transactions of the Royal Society of London Series B: Biological Sciences **357**:1153–1170.

Siddig, A. A. H., A. M. Ellison, A. Ochs, C. Villar-Leeman, and M. K. Lau. 2016. How do ecologists select and use indicator species to monitor ecological change? Insights from 14 years of publication in *Ecological Indicators*. Ecological Indicators **60**:223–230.

Sinervo, B., F. Méndez-de-la Cruz, D. B. Miles, B. Heulin, E. Bastiaans, M. Villagrán-Santa Cruz, et al. 2010. Erosion of lizard diversity by climate change and altered thermal niches. Science **328**:894–899.

Sirota, J., B. Baiser, N. J. Gotelli, and A. M. Ellison. 2013. Organic-matter loading determines regime shifts and alternative states in an aquatic ecosystem. Proceedings of the National Academy of Sciences of the USA **110**:7742–7747.

Smith, K. E., S. Thatje, and C. Hauton. 2013. Thermal tolerance during early ontogeny in the common whelk *Buccinum undatum* (Linneaus 1785): bioenergetics, nurse egg partitioning and developmental success. Journal of Sea Research **79**:32–39.

Somero, G. N., B. L. Lockwood, and L. Tomanek. 2017. Biochemical Adaptation: Response to Environmental Challenges from Life's Origin to the Anthropocene. Sinauer Associates, Sunderland, MA.

Sontag, S., 1969. "Thinking against oneself": Reflections on Cioran. In S. Fitzgerald and R. Fitzgerald, editors. Styles of Radical Will, pp. 74–95. Farrar, Straus & Giroux, New York.

Sork, V. L., F. W. Davis, R. Westfall, A. Flint, M. Ikegami, H. F. Wang, and D. Grivet. 2010. Gene movement and genetic association with regional climate gradients in California valley oak (*Quercus lobata* Née) in the face of climate change. Molecular Ecology **19**:3806–3823.

Sornette, D. 2006. Critical Phenomena in Natural Sciences, 2nd ed. Springer-Verlag, Berlin.

Sprengel, C. 1828. Von den Substanzen der Ackerkrume und des Untergrundes. Journal für Technische und Ökonomische Chemie **2**:423–474.

Srivastava, D. S. 2006. Habitat structure, trophic structure and ecosystem function: interactive effects in a bromeliad-insect community. Oecologia **149**:493–504.

Srivastava, D. S., J. Kolasa, J. Bengtsson, A. González, S. P. Lawler, T. E. Miller, et al. 2004. Are natural microcosms useful model systems for ecology? Trends in Ecology & Evolution **19**:379–384.

Steffen, W., K. Richardson, J. Rockström, S. E. Cornell, I. Fetzer, E. M. Bennett, et al. 2015. Planetary boundaries: guiding human development on a changing planet. Science **347**:1259855.

Steffen, W., J. Rockström, K. Richardson, T. M. Lenton, C. Folke, D. Liverman, et al. 2018. Trajectories of the earth system in the Anthropocene. Proceedings of the National Academy of Sciences of the USA **115**:8252–8259.

Steinbauer, M., J. A. Grytnes, G. Jurasinski, A. Kulonen, J. Lenoir, H. Pauli, et al. 2018. Accelerated increase in plant species richness on mountain summits is linked to warming. Nature **556**:231–234.

Steiner, C. F., A. S. Schwaderer, H. Veronkia, C. A. Klausmeier, and E. Litchman. 2009. Periodically forced food-chain dynamics: model predictions and experimental validation. Ecology **90**:3099–3107.

Stephens, P. A., W. J. Sutherland, and R. P. Freckleton. 1999. What is the Allee effect? Oikos **87**:185–190.

Sterner, R., and J. Elser. 2002. Ecological Stoichiometry: The Biology of Elements from Molecules to the Rhizosphere. Princeton University Press, Princeton, NJ.

Stevens, P. F., 2020. Angiosperm Phylogeny Website. Version 14, July 2017 [and more or less continuously updated since]. http://www.mobot.org/MOBOT/research/APweb/.

Stone, L., and A. Roberts. 1990. The checkerboard score and species distributions. Oecologia **85**:74–79.

Strona, G., D. Nappo, F. Boccacci, S. Fattorini, and J. San-Miguel-Ayanz. 2014. A fast and unbiased procedure to randomize ecological binary matrices with fixed row and column totals. Nature Communications **5**:4114.

Strona, G., W. Ulrich, and N. J. Gotelli. 2018. Bi-dimensional null model analysis of presence-absence binary matrices. Ecology **99**:103–115.

Strong, D. R., J. H. Lawton, and R. Southwood, editors. 1984. Insects on Plants: Community Patterns and Mechanisms. Harvard University Press, Cambridge, MA.

Stroud, J. T., M. R. Bush, M. C. Ladd, R. J. Nowicki, A. A. Shantz, and J. Sweatman. 2015. Is a community still a community? Reviewing definitions of key terms in community ecology. Ecology and Evolution **5**:4757–4765.

Sturm, M., T. Douglas, C. Racine, and G. E. Liston. 2005. Changing snow and shrub conditions affect albedo with global implications. Journal of Geophysical Research **110**:G01004.

Suding, K. N., and R. J. Hobbs. 2009. Threshold models in restoration and conservation: a developing framework. Trends in Ecology & Evolution **25**:271–279.

Sunday, J. M., A. E. Bates, M. R. Kearney, R. K. Colwell, N. K. Dulvy, J. T. Longino, and R. B. Huey. 2014. Thermal-safety margins and the necessity of thermoregulatory behavior across latitude and elevation. Proceedings of the National Academy of Sciences of the USA **111**:5610–5615.

Sutherland, J. P. 1974. Multiple stable points in natural communities. The American Naturalist **108**:859–873.

Svenning, J.-C., and F. Skov. 2004. Limited filling of the potential range in European tree species. Ecology Letters **7**:565–573.

Swan, J. M. A., and A. M. Gill. 1970. The origins, spread, and consolidation of a floating bog in Harvard Pond, Petersham, Massachusetts. Ecology **51**:829–840.

Szyrmer, J., and R. E. Ulanowicz. 1987. Total flows in ecosystems. Ecological Modelling **35**:123–136.

Talbot, C., R. Farhadi, and P. Aftabi. 2009. Potash in salt extruded at Sar Pohl diapir, Southern Iran. Ore Geology Reviews **35**:352–366.

Tanner, J. E., T. P. Hughes, and J. H. Connell. 1996. The role of history in community dynamics: a modelling approach. Ecology **77**:108–117.

Tansley, A. G. 1935. The use and abuse of vegetational terms and concepts. Ecology **16**:284–307.

ter Braak, C. J. F., and I. C. Prentice. 1988. A theory of gradient analysis. Advances in Ecological Research **18**:271–317.

Thom, D., W. Rammer, T. Dirnboeck, J. Mueller, J. Kobler, K. Katzensteiner, N. Helm, and R. Seidl. 2017. The impacts of climate change and disturbance on spatio-temporal trajectories of biodiversity in a temperate forest landscape. Journal of Applied Ecology **54**:28–38.

Thomas, K., and D. M. Cameron. 1986. Pollination and fertilization in the pitcher plant (*Sarracenia purpurea* L.). American Journal of Botany **73**:678.

Thompson, R. M., U. Brose, J. A. Dunner, R. O. J. Hall, S. Hladyz, R. L. Kitching, et al. 2012. Food webs: reconciling the structure and function of biodiversity. Trends in Ecology & Evolution **27**:689.

Thompson, R. M., and C. R. Townsend. 2005. Energy availability, spatial heterogeneity and ecosystem size predict food-web structure in streams. Oikos **108**:137–148.

Thorén, H., and L. Olsson. 2017. Is resilience a normative concept? Resilience: International Policies, Practices and Discourses **6**:112–128.

Tilly, L. J. 1968. The structure and dynamics of Cone Spring. Ecological Monographs **38**:169–197.

Tilman, D. 1982. Resource Competition and Community Structure. Princeton University Press, Princeton, NJ.

Tilman, D., C. Balzer, J. Hill, and B. L. Befort. 2011. Global food demand and the sustainable intensification of agriculture. Proceedings of the National Academy of Sciences of the USA **108**:20260–20264.

Tipping, E., S. Benham, J. Boyle, P. Crow, J. Davies, U. Fischer, et al. 2014. Atmospheric deposition of phosphorus to land and freshwater. Environmental Science: Processes & Impacts **16**:1608–1617.

Tournefort, J. P. d. 1700. Josephi Pitton Tournefort Aquisextiensis, doctoris medici Parisiensis, Academiae regiae scientiarum socii, et in horto regio botanices professoris, Institutiones rei herbariae. E. Typographia Regia, Paris.

Townsend, A., and R. W. Howarth. 2010. Human acceleration of the global nitrogen cycle. Scientific American **302**:32–39.

Trisos, C. H., C. Merow, and A. L. Pigot. 2020. The projected timing of abrupt ecological disruption from climate change. Nature **580**:496–501.

Troia, M. J., and R. A. McManamay. 2016. Filling in the GAPS: evaluating completeness and coverage of open-access biodiversity databases in the United States. Ecology and Evolution **6**:4654–4669.

Troll, W. 1932. Morphologie der schildformigen Blätter. Planta **17**:153–314.

Troll, W. 1939. Vergleichende Morphologie der hoheren Pflanzen, Band 1. Vegetationsorgane, Teil 2. Gebruder Borntrager, Berlin.

Troudet, J., P. Grandcolas, A. Blin, R. Vignes-Lebbe, and F. Legendre. 2017. Taxonomic bias in biodiversity data and societal preferences. Scientific Reports **7**:9132.

Turchin, P., and A. D. Taylor. 1992. Complex dynamics in ecological time series. Ecology **73**:289–305.

Ulanowicz, R. E. 2004. Quantitative methods for ecological network analysis. Computational Biology and Chemistry **28**:321–339.

Ulanowicz, R. E., R. D. Holt, and M. Barfield. 2014. Limits on ecosystem trophic complexity: insights from ecological network analysis. Ecology Letters **17**:127–136.

Ulrich, W., and N. J. Gotelli. 2007a. Disentangling community patterns of nestedness and species co-occurrence. Oikos **116**:2053–2061.

Ulrich, W., and N. J. Gotelli. 2007b. Null model analysis of species nestedness patterns. Ecology **88**:1824–1831.

Ulrich, W., Y. Kubota, B. Kusumoto, A. Baselga, H. Tuomisto, and N. J. Gotelli. 2018. Species richness correlates of raw and standardized co-occurrence metrics. Global Ecology and Biogeography **27**:395–399.

Ureta, C., C. Martorell, A. P. Curevo-Robayo, M. A. C. Mandujano, and E. MArtinez-Meyer. 2018. Inferring space from time: on the relationship between demography and environmental suitability in the desert plant *O. rastrera*. PLoS ONE **13**:e0201543.

US Fish and Wildlife Service. 1988. Endangered and threatened wildlife and plants; determination of endangered status for *Sarracenia rubra* ssp. *jonesii* (Mountain Sweet Pitcher Plant). Federal Register **53**:38470–38474.

US Fish and Wildlife Service. 1989. Endangered and threatened wildlife and plants; determination of endangered status for *Sarracenia rubra* ssp. *alabamensis* (Alabama Canebrake Pitcher Plant). Federal Register **54**:10150–10154.

Valiente-Banuet, A., M. Aizen, J. Alcántara, J. Arroyo, M. Coccuci, A., Galetti, et al. 2015. Beyond species loss: the extinction of ecological interactions in a changing world. Functional Ecology **29**:299–307.

Valls, A., M. Coll, and V. Christensen. 2015. Keystone species: toward an operational concept for marine biodiversity conservation. Ecological Monographs **85**:29–47.

van der Plas, F. 2019. Biodiversity and ecosystem functioning in naturally assembled communities. Biological Reviews **94**:1220–1245.

van der Ploeg, R. R., W. Böhm, and M. B. Kirkham. 1999. On the origin of the theory of mineral nutrition of plants and the Law of the Minimum. Soil Science Society of America Journal **63**:1055–1062.

van Nes, E. H., B. M. S. Arani, A. Staal, B. van der Bolt, B. M. Flores, S. Bathiany, and M. Scheffer. 2016. What do you mean, 'tipping point'? Trends in Ecology & Evolution **31**:902–904.

Van Straalen, N. I., and D. Roelofs. 2012. Introduction to Ecological Genomics. Oxford University Press, Oxford.

van Tienderen, P. 2000. Elasticities and the link between demographic and evolutionary dynamics. Ecology **81**:666–679.

Vandenkoornhuyse, P., A. Dufresne, A. Quaiser, G. Gouesbet, F. Binet, A.-J. Francez, et al. 2010. Integration of molecular functions at the ecosystemic level: breakthroughs and future goals of environmental genomics and post-genomics. Ecology Letters **13**:776–791.

Vaughan, I. P., and N. J. Gotelli. 2019. Water quality improvements offset the climatic debt for stream macroinvertebrates over twenty years. Nature Communications **10**:1956.

Veraart, A. J., E. J. Faassen, V. Dakos, E. H. van Nes, M. Lürling, and M. Scheffer. 2012. Recovery rates reflect distance to a tipping point in a living system. Nature **481**:357–359.

Violle, C., B. J. Enquist, B. J. McGill, L. Jiang, C. H. Albert, C. Hulshof, V. Jung, and J. Messier. 2012. The return of the variance: intraspecific variability in community ecology. Trends in Ecology & Evolution **27**:244–252.

Vitousek, P. M. 1994. Beyond global warming: ecology and global change. Ecology **75**:1861–1876.

Vogel, C., and E. M. Marcotte. 2012. Insights into the regulation of protein abundance from proteomic and transcriptomic analyses. Nature Reviews Genetics **13**:227–232.

Vogel, S. 1998. Remarkable nectaries: structure, ecology, organophyletic perspectives. II. Nectarioles. Flora **193**:1–29.

Voltaire. 1957 edition. Candide; ou, L'optimisme. Librairie M. Didier, Paris.

Wagg, C., S. F. Bender, F. Widmer, and M. G. A. van der Heijden. 2014. Soil biodiversity and soil community composition determine ecosystem multifunctionality. Proceedings of the National Academy of Sciences of the USA **111**:5266–5270.

Wakefield, A. E., N. J. Gotelli, S. E. Wittman, and A. M. Ellison. 2005. Prey addition alters nutrient stoichiometry of the carnivorous plant *Sarracenia purpurea*. Ecology **86**:1737–1743.

Walker, B. H. 1992. Biodiversity and ecological redundancy. Conservation Biology **6**:18–23.

Walker, K. J. 2014. *Sarracenia purpurea* subsp. *purpurea* (Sarraceniaceae) naturalised in Britain and Ireland: distribution, ecology, impacts and control. New Journal of Botany **4**:33–41.

Walker, K. J., C. Auld, E. Austin, and J. Rook. 2016. Effectiveness of methods to control the invasive non-native pitcherplant *Sarracenia purpurea* L. on a European mire. Journal for Nature Conservation **31**:1–8.

Walther, G.-R., E. Post, P. Convey, A. Menzel, C. Parmesan, T. J. C. Beebee, et al. 2002. Ecological responses to recent climate change. Nature **416**:389–395.

Ware, I. M., C. R. Fitzpatrick, A. Senthilnathan, S. L. J. Bayliss, K. K. Beals, L. O. Mueller, et al. 2019. Feedbacks link ecosystem ecology and evolution across spatial and temporal scales: empirical evidence and future directions. Functional Ecology **33**: 31–42.

Warming, E. 1895. Plantesamfund: Grundtræk af den økologiske Plantegeografi. P.G. Philipsens Forlag, Copenhagen.

Warnes, G. R., B. Bolker, and T. Lumley. 2018. gtools: Various R Programming Tools. https://cran.r-project.org/package=gtools.

Warren, D., C. Kraft, D. Josephson, and C. Driscoll. 2017. Acid rain recovery may help to mitigate the impacts of climate change on thermally sensitive fish in lakes across eastern North America. Global Change Biology **23**:2149–2153.

Waser, N. M., R. K. Vickery, and M. V. Price. 1982. Patterns of seed dispersal and population differentiation in *Mimulus guttatus*. Evolution **36**:753–761.

Watanabe, T., M. R. Broadley, S. Jansen, P. J. White, J. Takada, K. Satake, T. Takamatsu, S. J. Tuah, and O. M. 2007. Evolutionary control of leaf element composition in plants. New Phytologist **174**:516–523.

Watts, D. J., and S. H. Strogatz. 1998. Collective dynamics of 'small-world' networks. Nature **393**:440–442.

Weiher, E., and P. Keddy, editors. 1999. Ecological Assembly Rules: Perspectives, Advances, Retreats. Cambridge University Press, Cambridge.

Weisberg, M. 2013. Simulation and Similarity: Using Models to Understand the World. Oxford University Press, Oxford.

West, G. B. 2017. Scale: The Universal Laws of Growth, Innovation, Sustainability and the Pace of Life in Organisms, Cities, Economies, and Companies. Penguin Press, New York.

West, G. B., J. H. Brown, and B. J. Enquist. 1997. A general model for the origin of allometric scaling laws in biology. Science **276**:122–126.

Westerling, A. L., A. Gershunov, T. J. Brown, D. R. Cayan, and M. D. Dettinger. 2003. Climate and wildfire in the western United States. Bulletin of the American Meteorological Society **84**:595–604.

Weterings, R., C. Umponstira, and H. L. Buckley. 2018. Landscape variation influences trophic cascades in dengue vector food webs. Science Advances **4**:eaap9534.

Wherry, E. T. 1933. The geographic relations of *Sarracenia purpurea*. Bartonia **15**:1–6.

White, J., K. Nickols, D. Malone, M. Carr, R. Starr, F. Cordoleani, et al. 2016. Fitting state-space integral projection models to size-structured time series data to estimate unknown parameters. Ecological Applications **26**:2675–2692.

Whitewoods, C. D., B. Gonçalves, J. Cheng, M. Cui, R. Kennaway, K. Lee, et al. 2020. Evolution of carnivorous traps from planar leaves through simple shifts in gene expression. Science **367**:91–96.

Whitham, T. G., J. K. Bailey, J. A. Schweitzer, S. M. Shuster, R. K. Bangert, C. J. LeRoy, et al. 2006. A framework for community and ecosystem genetics: from genes to ecosystems. Nature Reviews Genetics **7**:510–523.

Whittaker, R. H. 1935. Vegetation of the Great Smoky Mountains. Ecology **26**:1–80.

Whittaker, R. H. 1962. Classification of natural communities. Botanical Review **28**:1–239.

Whittaker, R. J., M. B. Bush, and K. Richards. 1989. Plant recolonization and vegetation succession on the Krakatau islands, Indonesia. Ecological Monographs **59**: 59–123.

Whyndham, J. 1951. The Day of the Triffids. Michael Joseph, London.

Wiens, J., N. H. Schumaker, R. Inman, T. Esque, K. Longshore, and K. Nussear. 2017. Spatial demographic models to inform conservation planning of Golden Eagles in renewable energy landscapes. Journal of Raptor Research **51**:234–257.

Wiens, J. A. 1989. Spatial scaling in ecology. Functional Ecology **3**:385–397.

Williams, G. P. 1989. Sediment concentration versus water discharge during single hydrologic events in rivers. Journal of Hydrology **111**:89–106.

Williams, K., C. B. Field, and H. A. Mooney. 1989. Relationships among leaf construction cost, leaf longevity, and light environment in rain-forest plants of the genus *Piper*. The American Naturalist **133**:198–211.

Willig, M. R., and L. R. Walker, editors. 2016. Long-Term Ecological Research: Changing the Nature of Scientists. Oxford University Press, New York.

Wilmes, P., and P. L. Bond. 2006. Metaproteomics: studying functional gene expression in microbial ecosystems. Trends in Microbiology **14**:92–97.

With, K. A., and T. O. Crist. 1995. Critical thresholds in species' responses to landscape structure. Ecology **76**:2446–2459.

Wittebolle, L., M. Marzorati, L. Clement, A. Balloi, D. Daffonchio, K. Heylen, et al. 2009. Initial community evenness favours functionality under selective stress. Nature **458**:623–626.

Wolfe, L. M. 1981. Feeding behavior of a plant: differential prey capture in old and new leaves of the pitcher plant (*Sarracenia purpurea*). The American Midland Naturalist **106**:352–359.

Wooton, J. T. 1994. The nature and consequences of indirect effects in ecological communities. Annual Review of Ecology and Systematics **24**: 443–466.

Wootton, J. T. 1998. Effects of disturbance on species diversity: a multi-trophic perspective. The American Naturalist **152**:803–825.

Wootton, J. T. 2010. Experimental species removal alters ecological dynamics in a natural ecosystem. Ecology **91**:42–48.

Wray, D. L., and C. S. Brimley. 1943. The insect inquilines and victims of pitcher plants in North Carolina. Annals of the Entomological Society of America **36**:128–137.

Wright, I. J., P. B. Reich, J. H. C. Cornelissen, D. S. Falster, E. Garnier, K. Hikosaka, et al. 2005. Assessing the generality of global leaf trait relationships. New Phytologist **166**:485–496.

Wright, I. J., P. B. Reich, M. Westoby, D. D. Ackerly, Z. Baruch, F. Bongers, et al. 2004. The worldwide leaf economics spectrum. Nature **428**:821–827.

Wright, K. W., K. L. Vanderbilt, D. W. Inouye, C. D. Bertelsen, and T. M. Crimmins. 2015. Turnover and reliability of flower communities in extreme environments: insights from long-term phenology data sets. Journal of Arid Environments **115**:27–34.

Yackulic, C. B., R. Chandler, E. F. Zipkin, J. A. Royle, J. D. Nichols, E. H. Campbell Grant, and S. Veran. 2013. Presence-only modelling using MAXENT: when can we trust the inferences? Methods in Ecology and Evolution **4**:236–243.

Yasuhara, M., T. M. Cronin, P. B. deMenocal, H. Okahashi, and B. K. Linsley. 2008. Abrupt climate change and collapse of deep-sea ecosystems. Proceedings of the National Academy of Sciences of the USA **105**:1556–1560.

Yeakel, J. D., M. Pires, L. Rudolf, N. J. Dominy, P. L. Koch, P. R. Guimarães, and T. Gross. 2014. Collapse of an ecological network in Ancient Egypt. Proceedings of the National Academy of Sciences of the USA **111**:14472–14477.

Yoshimura, J., K. Tainaka, T. Suzuki, Y. Sakisaka, N. Nakagiri, T. Togashi, and T. Miyazaki. 2006. The role of rare species in the community stability of a model ecosystem. Evolutionary Ecology Research **8**:629–642.

Yu, Z., S. M. B. Krause, D. A. C. Beck, and L. Chistoserdova. 2016. A synthetic ecology perspective: how well does behavior of model organisms in the laboratory predict microbial activities in natural habitats? Frontiers in Microbiology **7**:946.

Zajac, R. N., and R. B. Whitlach. 1982. Responses of estuarine infauna to disturbance. 2. Spatial and temporal variation of succession. Marine Ecology Progress Series **10**:15–27.

Zajac, R. N., R. B. Whitlach, and S. F. Thrush. 1998. Recolonization and succession in soft-sediment infaunal communities: the spatial scale of controlling factors. Hydrobiologia **375-376**:227–240.

Zaman, A., and D. Simberloff. 2002. Random binary matrices in biogeographical ecology: Instituting a good neighbor policy. Environmental and Ecological Statistics **9**:405–421.

Zaoli, S., A. Giometto, A. Maritan, and A. Rinaldo. 2017. Covariations in ecological scaling laws fostered by community dynamics. Proceedings of the National Academy of Sciences of the USA **114**:10672–10677.

Zeileis, A., C. Kleiber, W. Krämer, and K. Hornik. 2003. Testing and dating of structural changes in practice. Computational Statistics and Data Analysis **44**:109–123.

Zhang, J., X. Li, X. Wang, W.-X. Wang, and L. Wu. 2015. Scaling behaviours in the growth of networked systems and their geometric origins. Scientific Reports **5**:9767.

Zhang, Q., and A. D. Werner. 2015. Hysteretic relationships in inundation dynamics for a large lake-floodplain system. Journal of Hydrology **527**:160–171.

Zhang, S., S. R. Chaluvadi, and J. L. Bennetzen. 2020a. Draft genome sequence of a *Enterobacter* sp. C6, found in the pitcher fluids of *Sarracenia rosea*. Microbiology Resource Announcements **9**:e01214–19.

Zhang, S., S. R. Chaluvadi, and J. L. Bennetzen. 2020b. Draft genome sequence of a *Serratia marcescens* strain isolated from the pitcher fluids of a *Sarracenia* pitcher plant. Microbiology Resource Announcements **9**:e01216–19.

Zohner, C. M., A. Rockinger, and S. S. Renner. 2019. Increased autumn productivity permits temperate trees to compensate for spring frost damage. New Phytologist **221**:789–795.

Zor, T., and Z. Seliger. 1996. Linearization of the Bradford protein assay increases its sensitivity: theoretical and experimental studies. Analytical Biochemistry **236**:302–308.

Zotz, G., and R. Asshoff. 2010. Growth in epiphytic bromeliads: response to the relative supply of phosphorus and nitrogen. Plant Biology **12**:108–113.

Subject Index

Taxonomic Index

MONOGRAPHS IN POPULATION BIOLOGY

SIMON A. LEVIN, ROBERT PRINGLE, AND CORINA TARNITA, SERIES EDITORS

Ingram Content Group UK Ltd.
Milton Keynes UK
UKHW021042190523
422012UK00006B/201